Number
CONNECTIONS

D1808679

second edition

Copymasters

Yellow Level

ROSE GRIFFITHS

Heinemann

Heinemann Educational Publishers
Halley Court, Jordan Hill, Oxford, OX2 8EJ
a division of Harcourt Education Ltd
www.myprimary.co.uk

Heinemann is a registered trademark of Harcourt Education Ltd

First edition first published 1996

Second edition first published 2005

10 09 08 07 06 05
10 9 8 7 6 5 4 3 2 1

ISBN 0 435 02170 2

Designed and typeset by Susan Clarke
Illustrated by Teri Gower and Jeff Edwards
Cover design by Susan Clarke
Repro by Digital Imaging, Glasgow
Printed and bound in the UK by Thomson Litho

The author and publishers would like to thank teachers at the following schools for their help in trialling these materials:
Folville Junior School, Leicester
Wolsey House Primary School, Leicester
Emmer Green Primary School, Reading
St Anne's Primary School, Streetly
Orton Wistow Primary School, Peterborough
Spooner Row Primary School, Wymondham
Stafford Leys Primary School, Leicester
St Peter's Primary School, Blaenavon

Contents

Y1

Seeds and shells

Name ______________________

Date ______________________

Dessertspoon

DRIED PEAS

PAPER CLIPS

Do as many of these as you can.

About how many do you think you would get in a spoonful? Count to see.

Things to count	Estimate of a spoonful	Actual spoonful
Paperclips		
Centicubes		
Counters		
Pennies		
Tags		
Split pins		
Dried peas		
Macaroni		

Months and years

Name ____________________

Date ____________________

Fill in the missing letters.

March	April	May
M ______	A ______	M ____
__________	__________	________

January	February	August
Jan ______	Febr ______	Aug _____
J __________	F __________	Aug _____
__________	__________	______ust
		______ust
		A ________
		__________

June	July
J _____	J _____
______	______

September	November	December
Sept ________	Nov ________	Dec ________
S __________	N __________	D __________
__________	__________	__________

October Oct _______ O ____________

Writing dates

◄ Yellow Pupil Book Part 1 pages 10 and 11
► Copymaster Y3

Months and years

Name ____________________

Date ____________________

Y3

Fill in the missing years.

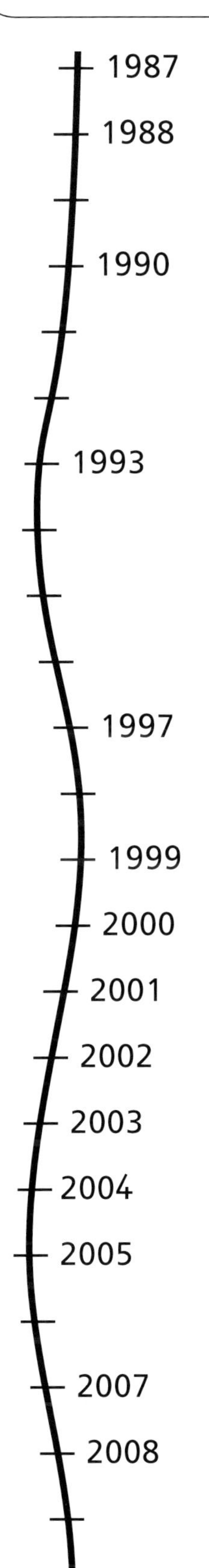
1987
1988
1990
1993
1997
1999
2000
2001
2002
2003
2004
2005
2007
2008
2010

Write each date the longer way.

1 / 2 / 93	1st February 1993
17 / 5 / 87	____________
14 / 3 / 10	____________
30 / 8 / 05	____________
24 / 10 / 00	____________
23 / 1 / 92	____________
2 / 11 / 02	____________
10 / 7 / 09	____________
9 / 4 / 01	____________

Jim was born in 1997.

Tom was born in 2000.

Sara was born in 1998.

Who is oldest? ____________

Owen was born in 1999.

Kate was born in 2001.

Ranvir was born in 2004.

Who is oldest? ____________

◄ Yellow Pupil Book Part 1 pages 10 and 11

Off by heart

Name ____________________

Date ____________________

Y4

Fill in the missing numbers.
Check on a tables square. ✓ or ✗

$\square \times 4 = 8$	$\square \times 2 = 4$	$\square \times 3 = 9$
$\square \times 4 = 16$	$\square \times 2 = 0$	$\square \times 3 = 3$
$\square \times 4 = 4$	$\square \times 2 = 10$	$\square \times 3 = 12$
$\square \times 4 = 12$	$\square \times 2 = 2$	$\square \times 3 = 0$
$\square \times 4 = 20$	$\square \times 2 = 6$	$\square \times 3 = 15$
$\square \times 4 = 0$	$\square \times 2 = 8$	$\square \times 3 = 6$

$25 \div 5 =$ ____	$8 \div 2 =$ ____	$10 \div 2 =$ ____
$12 \div 4 =$ ____	$15 \div 5 =$ ____	$16 \div 4 =$ ____
$20 \div 5 =$ ____	$12 \div 3 =$ ____	$5 \div 5 =$ ____
$6 \div 3 =$ ____	$8 \div 4 =$ ____	$10 \div 5 =$ ____
$20 \div 4 =$ ____	$4 \div 4 =$ ____	$15 \div 3 =$ ____

Ways of adding

Name ____________________

Date ____________________

Y13

Do these in your head.

62 + 10 = ____

20 + 47 = ____

20 + 60 = ____

30 + 29 = ____

51 + 20 = ____

55 + 20 = ____

44 + 30 = ____

50 + 22 = ____

30 + 40 = ____

30 + 33 = ____

✓ or ✗

Do these with a calculator.

47 + 18 = ____

29 + 18 = ____

58 + 18 = ____

42 + 18 = ____

21 + 18 = ____

35 + 18 = ____

36 + 18 = ____

51 + 18 = ____

☐ + 18 = 46

☐ + 18 = 67

Y14

Ways of adding

Name ______________________

Date ______________________

Work with a friend. Talk about how you will do each sum.

In your head? On a calculator? With tens and ones? On paper?

Do each sum, and write how you did it.

Forty add twenty-two

Fifty-six add nineteen

Twenty-six add thirty-seven

Thirty add fifteen

Eighteen add fifty-five

Forty-nine add ten

Times 2, times 3

Name ____________________

Date ____________________

Use tens and ones.

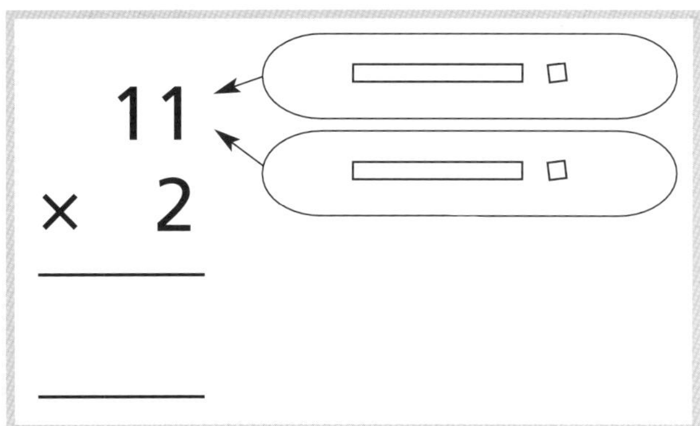

$$\begin{array}{r} 11 \\ \times \quad 2 \\ \hline \\ \hline \end{array}$$

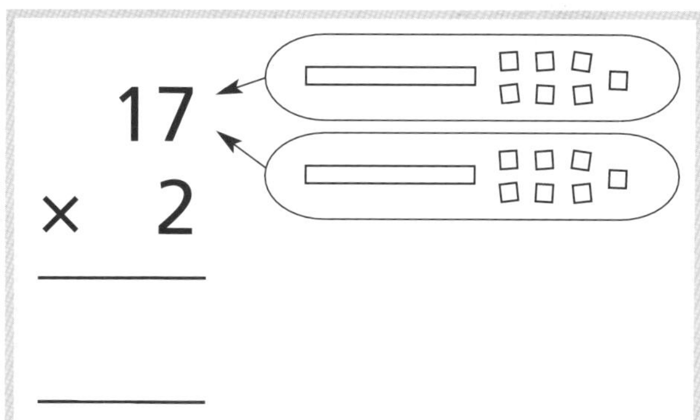

$$\begin{array}{r} 17 \\ \times \quad 2 \\ \hline \\ \hline \end{array}$$

Draw tens and ones.

$$\begin{array}{r} 13 \\ \times \quad 3 \\ \hline \\ \hline \end{array}$$

$$\begin{array}{r} 22 \\ \times \quad 2 \\ \hline \\ \hline \end{array}$$

$$\begin{array}{r} 12 \\ \times \quad 2 \\ \hline \\ \hline \end{array}$$

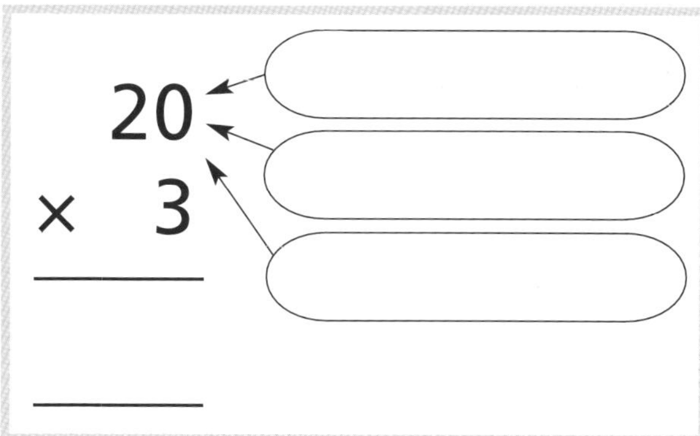

$$\begin{array}{r} 20 \\ \times \quad 3 \\ \hline \\ \hline \end{array}$$

$$\begin{array}{r} 18 \\ \times \quad 2 \\ \hline \\ \hline \end{array}$$

$$\begin{array}{r} 21 \\ \times \quad 3 \\ \hline \\ \hline \end{array}$$

Y16

Times 2, times 3

Name ____________________

Date ____________________

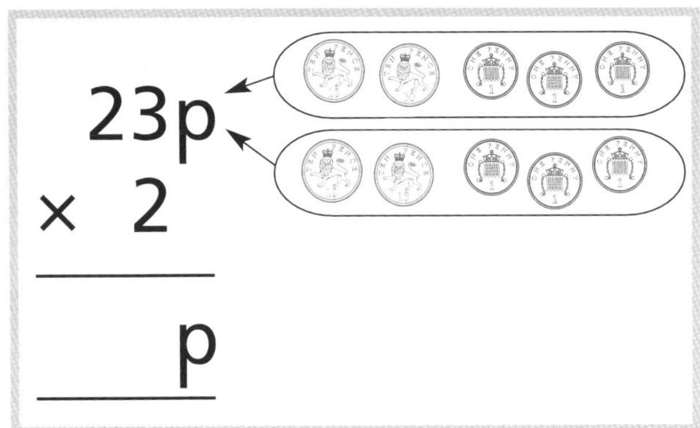
Use 10p and 1p coins.

23p
× 2
______
____p

15p
× 2
______
______

34p
× 2
______
______

17p
× 2
______
______

21p
× 3
______
______

21p
21p
+ 21p
______
______

32p
× 2
______
______

32p
+ 32p
______
______

Adding nine

Name ____________________

Date ____________________

Fill in the
missing numbers.

23 + 10 = ____
23 + 9 = ____

32 + 10 = ____
32 + 9 = ____

47 + 10 = ____
47 + 9 = ____

64 + 10 = ____
64 + 9 = ____

16 + 10 = ____
16 + 9 = ____

55 + 10 = ____
55 + 9 = ____

21 + 10 = ____
21 + 9 = ____

48 + 10 = ____
48 + 9 = ____

These are
easy sums.

20 + 9 = ____

30 + 9 = ____

40 + 9 = ____

50 + 9 = ____

Adding nine

Name ______________

Date ______________

Fill in the missing numbers.

46 + ☐ = 56

46 + 9 = ____

57 + 10 = ____

57 + 9 = ____

31 + 10 = ____

31 + ☐ = 40

70 + 10 = ____

70 + ☐ = 79

29 + 9 = ____

43 + 9 = ____

63 + 9 = ____

34 + 9 = ____

52 + 9 = ____

17 + 9 = ____

61 + 9 = ____

71 + 9 = ____

9 + 9 = ____

18 + 9 = ____

27 + 9 = ____

36 + 9 = ____

45 + 9 = ____

54 + 9 = ____

68 + 9 = ____

72 + 9 = ____

◀ Yellow Pupil Book Part 1 pages 26 and 27

More multiplying

Name ______________________

Date ______________________

Use tens and ones. Draw them.
Then multiply on paper.

15 times 2

15 times 3

15 times 4

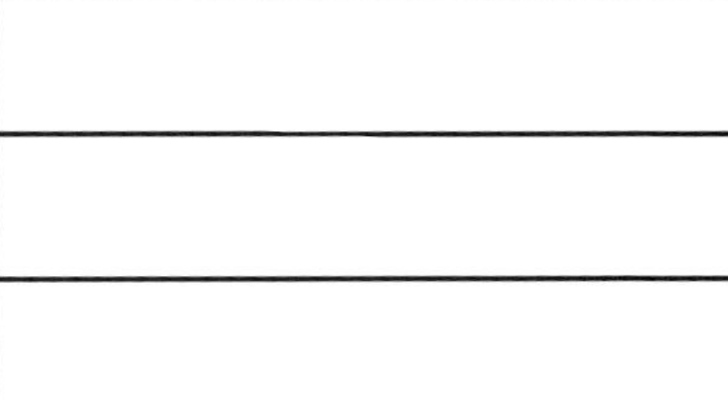

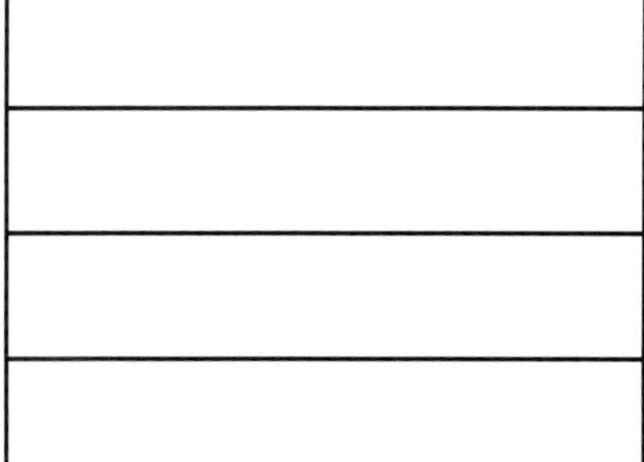

$$\begin{array}{r} 15 \\ \times\ \ 2 \\ \hline \\ \hline \end{array}$$

$$\begin{array}{r} 15 \\ \times\ \ 3 \\ \hline \\ \hline \end{array}$$

$$\begin{array}{r} 15 \\ \times\ \ 4 \\ \hline \\ \hline \end{array}$$

17 times 2

17 times 3

17 times 4

$$\begin{array}{r} 17 \\ \times\ \ 2 \\ \hline \\ \hline \end{array}$$

$$\begin{array}{r} 17 \\ \times\ \ 3 \\ \hline \\ \hline \end{array}$$

$$\begin{array}{r} 17 \\ \times\ \ 4 \\ \hline \\ \hline \end{array}$$

More multiplying

Name ____________________

Date ____________________

Use tens and ones. Draw them. Then multiply on paper.

16 times 2

16 times 3

16 times 4

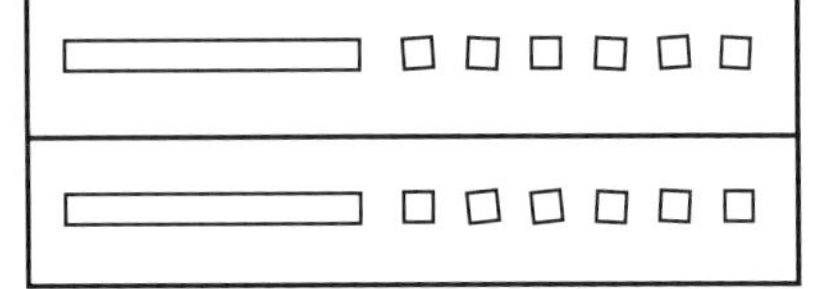

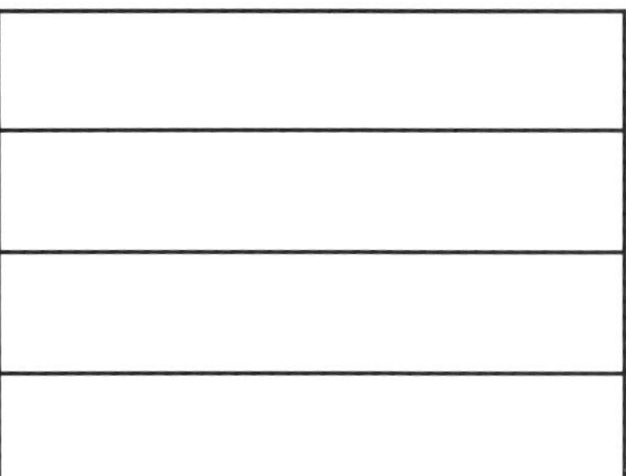

$$\begin{array}{r} 16 \\ \times \quad 2 \\ \hline \\ \hline \end{array} \qquad \begin{array}{r} 16 \\ \times \quad 3 \\ \hline \\ \hline \end{array} \qquad \begin{array}{r} 16 \\ \times \quad 4 \\ \hline \\ \hline \end{array}$$

Use tens and ones. Then multiply on paper.

$$\begin{array}{r} 33 \\ \times \quad 2 \\ \hline \\ \hline \end{array} \qquad \begin{array}{r} 37 \\ \times \quad 2 \\ \hline \\ \hline \end{array} \qquad \begin{array}{r} 14 \\ \times \quad 4 \\ \hline \\ \hline \end{array} \qquad \begin{array}{r} 14 \\ \times \quad 3 \\ \hline \\ \hline \end{array}$$

$$\begin{array}{r} 11 \\ \times \quad 4 \\ \hline \\ \hline \end{array} \qquad \begin{array}{r} 20 \\ \times \quad 4 \\ \hline \\ \hline \end{array} \qquad \begin{array}{r} 26 \\ \times \quad 3 \\ \hline \\ \hline \end{array}$$

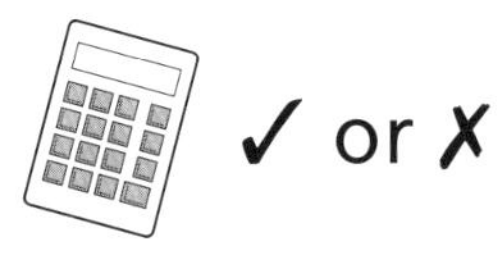

✓ or ✗

Add or take away

Name ______________________

Date ______________________

Y5

Write each sum and work it out.

Forty-four add thirty-one

$$\begin{array}{r} 44 \\ +\,31 \\ \hline \\ \hline \end{array}$$

Fifty-seven add six

Sixty-one take away nineteen

Eighty take away fifty-five

Twenty-eight add forty-eight

Twenty-five add fifty-three

Seventy-eight take away fifty-nine

Seventy-three take away forty-two

Add or take away

Name ____________________

Date ____________________

Use tens and ones.
Then take away on paper.

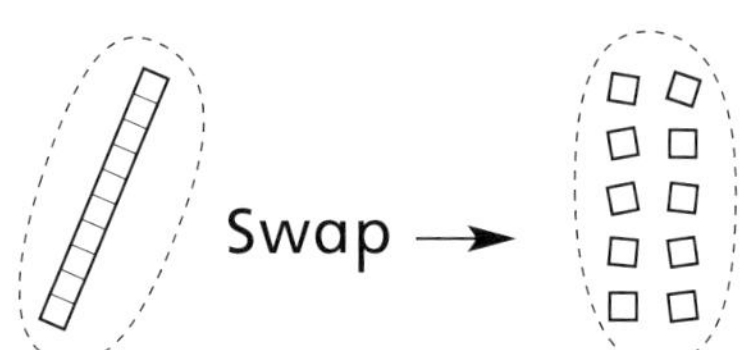
Swap

80 − 37 = ___	80 − 18 = ___	80 − 56 = ___	80 − 45 = ___
80 − 21 = ___	80 − 62 = ___	80 − 80 = ___	80 − 74 = ___

Fill in + or − .

36 ☐ 23 = 59   78 ☐ 15 = 63

45 ☐ 11 = 34   80 ☐ 78 = 2

53 + 26 = ___	8 + 42 = ___	48 + 28 = ___	25 + 47 = ___
65 − 27 = ___	39 − 18 = ___	73 − 14 = ___	70 − 40 = ___

Days and dates

Name ______________________

Date ______________________

How many days are there in July? __________

Fill in the missing numbers.

July 2004

Mon.	Tues.	Wed.	Thurs.	Fri.	Sat.	Sun.
			1	2		
5	6	7			10	11
12	13		15			18
			22			
				30		

In different years, July starts on different days.
Fill in the missing numbers.

July 2008

Mon.	Tues.	Wed.	Thurs.	Fri.	Sat.	Sun.
	1					

Days and dates

Name ____________________

Date ____________________

How many days in each month?

Fill in the missing numbers.

January	31
February	28 or 29
March	
April	

May	
June	
July	
August	

September	
October	
November	
December	

How many days in February this year? __________

Is this year a leap year? __________

Cut out these months, and put them in the right places on Copymaster Y9.

Think
How many days in each month?

Which day should each month start on?

M	T	W	T	F	S	S
1	2	3	4	5	6	7
8	9	10	11	12	13	14
15	16	17	18	19	20	21
22	23	24	25	26	27	28
29	30					

M	T	W	T	F	S	S
			1	2	3	4
5	6	7	8	9	10	11
12	13	14	15	16	17	18
19	20	21	22	23	24	25
26	27	28	29	30		

M	T	W	T	F	S	S
	1	2	3	4	5	6
7	8	9	10	11	12	13
14	15	16	17	18	19	20
21	22	23	24	25	26	27
28	29	30				

M	T	W	T	F	S	S
1	2	3	4	5	6	7
8	9	10	11	12	13	14
15	16	17	18	19	20	21
22	23	24	25	26	27	28

M	T	W	T	F	S	S
		1	2	3	4	5
6	7	8	9	10	11	12
13	14	15	16	17	18	19
20	21	22	23	24	25	26
27	28	29	30			

◀ Yellow Pupil Book Part 1 pages 16 and 17
▶ Copymaster Y9

Days and dates

Name ______________________

Date ______________________

Y9

Use with Copymaster Y8.

1999

January

M	T	W	T	F	S	S
				1	2	3
4	5	6	7	8	9	10
11	12	13	14	15	16	17
18	19	20	21	22	23	24
25	26	27	28	29	30	31

February

March

M	T	W	T	F	S	S
1	2	3	4	5	6	7
8	9	10	11	12	13	14
15	16	17	18	19	20	21
22	23	24	25	26	27	28
29	30	31				

April

May

M	T	W	T	F	S	S
					1	2
3	4	5	6	7	8	9
10	11	12	13	14	15	16
17	18	19	20	21	22	23
24	25	26	27	28	29	30
31						

June

July

M	T	W	T	F	S	S
			1	2	3	4
5	6	7	8	9	10	11
12	13	14	15	16	17	18
19	20	21	22	23	24	25
26	27	28	29	30	31	

August

M	T	W	T	F	S	S
						1
2	3	4	5	6	7	8
9	10	11	12	13	14	15
16	17	18	19	20	21	22
23	24	25	26	27	28	29
30	31					

September

October

M	T	W	T	F	S	S
				1	2	3
4	5	6	7	8	9	10
11	12	13	14	15	16	17
18	19	20	21	22	23	24
25	26	27	28	29	30	31

November

December

M	T	W	T	F	S	S
		1	2	3	4	5
6	7	8	9	10	11	12
13	14	15	16	17	18	19
20	21	22	23	24	25	26
27	28	29	30	31		

Speedy tables A

1 2 3 minute test

Name ______________________

Date ______________________

2 × 10 = ____	25 ÷ 5 = ____	2 × 9 = ____
3 × 4 = ____	60 ÷ 10 = ____	20 ÷ 4 = ____
9 × 2 = ____	8 ÷ 2 = ____	6 × 2 = ____
0 × 5 = ____	15 ÷ 3 = ____	10 ÷ 5 = ____
2 × 7 = ____	12 ÷ 4 = ____	8 × 0 = ____
8 × 2 = ____	4 ÷ 1 = ____	9 ÷ 3 = ____
4 × 4 = ____	14 ÷ 2 = ____	 **Score:** ____

Mental recall of tables facts ◄ Yellow Textbook 1 pages 14 and 15 onwards

Speedy tables B

1 2 3 minute test

Name ______________________

Date ______________________

8 × 1 = ____	18 ÷ 2 = ____	4 × 4 = ____
4 × 10 = ____	10 ÷ 2 = ____	16 ÷ 2 = ____
3 × 5 = ____	80 ÷ 10 = ____	0 × 5 = ____
2 × 8 = ____	20 ÷ 5 = ____	12 ÷ 4 = ____
5 × 5 = ____	9 ÷ 9 = ____	5 × 4 = ____
2 × 4 = ____	6 ÷ 2 = ____	15 ÷ 5 = ____
7 × 2 = ____	9 ÷ 3 = ____	 **Score:** ____

Squares

Name ____________________

Date ____________________

Y11

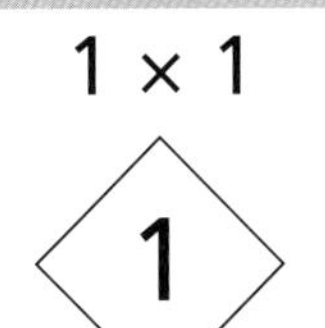
1 × 1
1

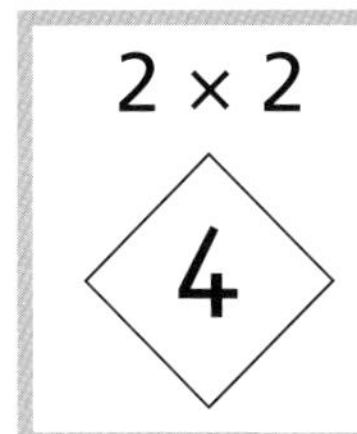
2 × 2
4

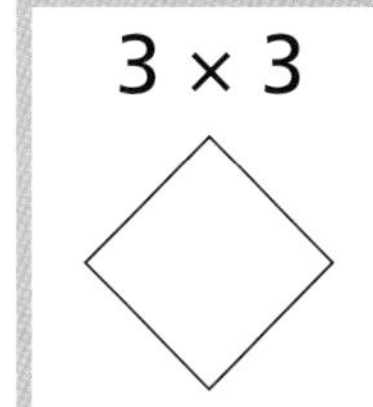
3 × 3

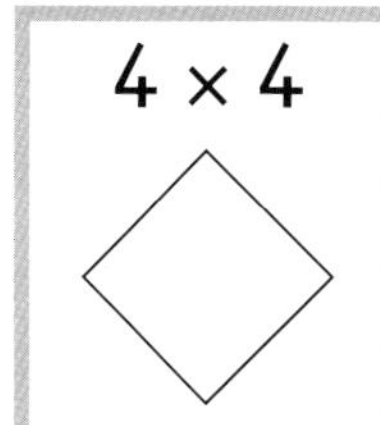
4 × 4

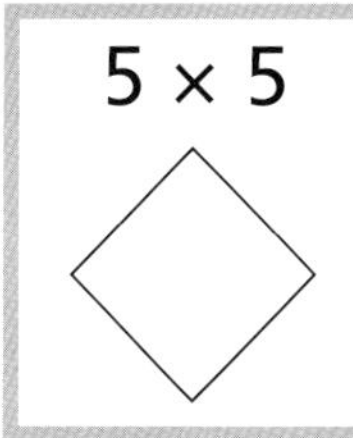
5 × 5

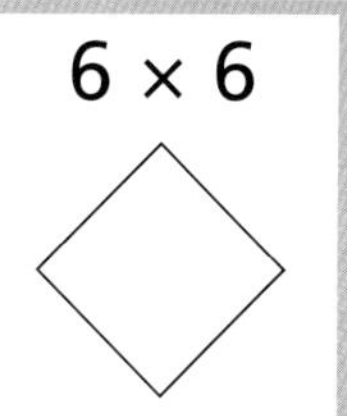
6 × 6

Fill in the missing square numbers.

Make square patterns with 25 bricks.
Choose your favourites and colour them here.

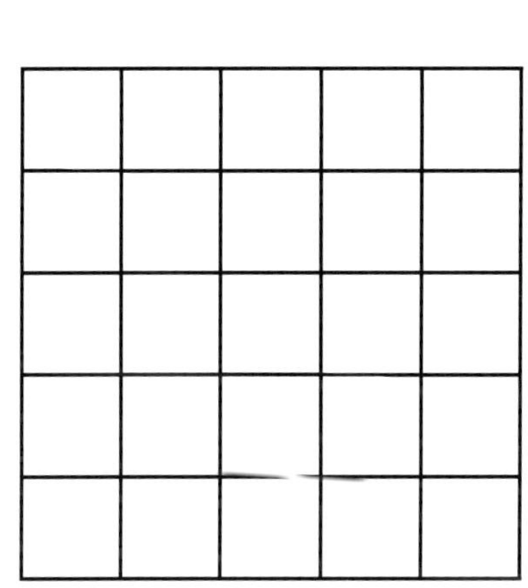

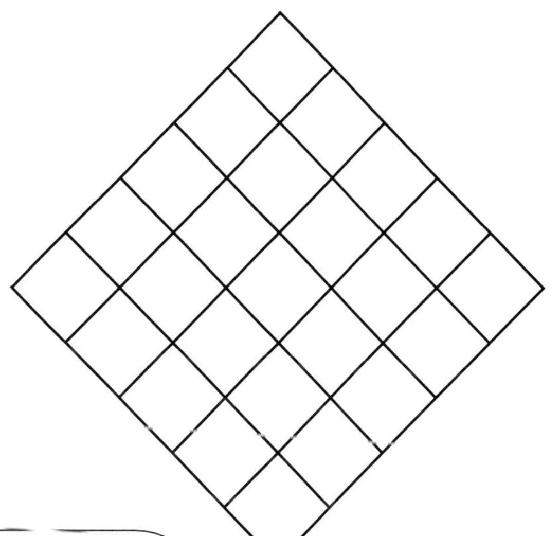

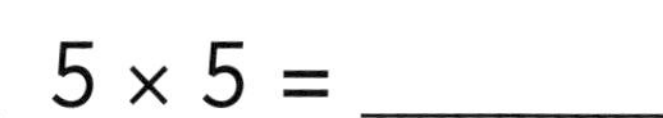
5 × 5 = ______

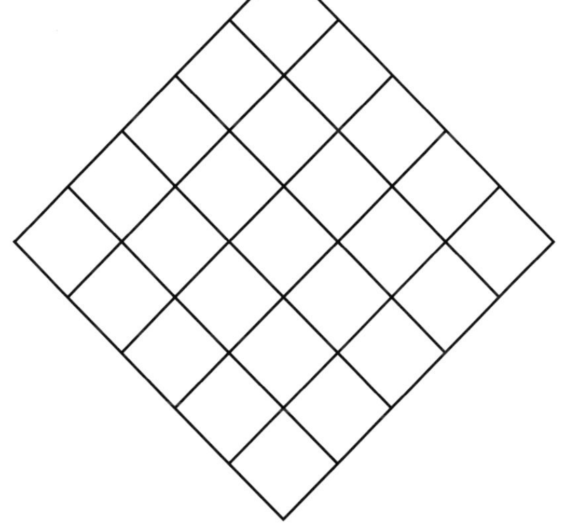

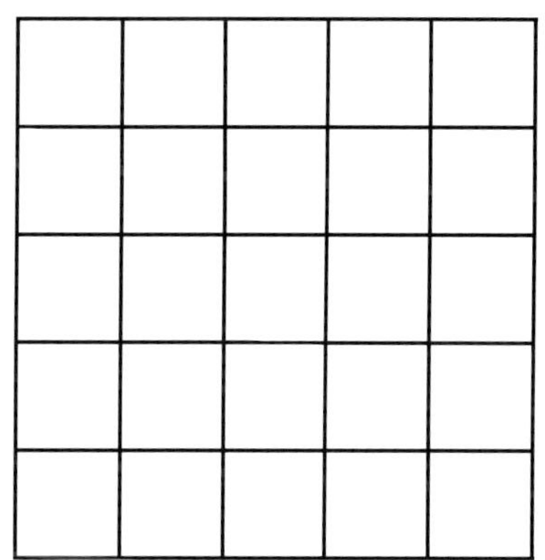

25 is a square number.

Squares

Name ______________________

Date ______________________

Make square patterns with 36 bricks.
Choose your favourites and colour them here.

6 × 6 = ________

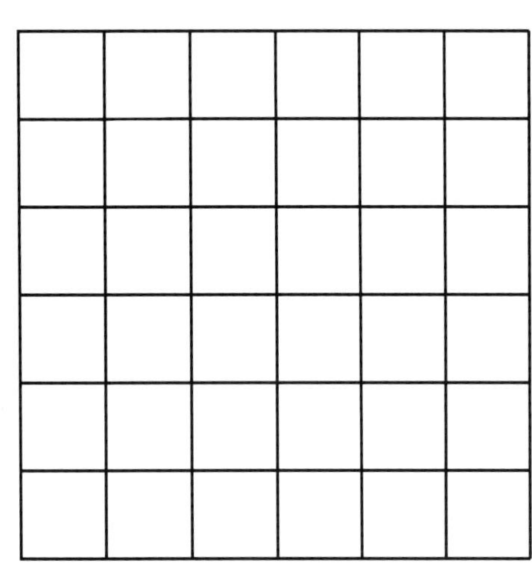

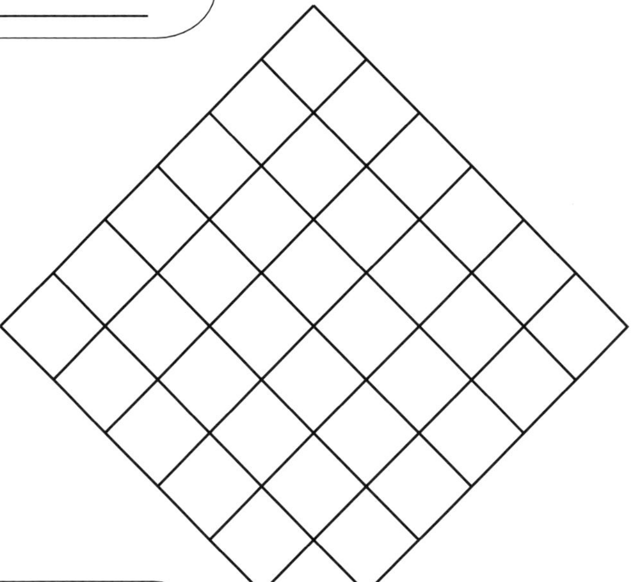

36 is a square number.

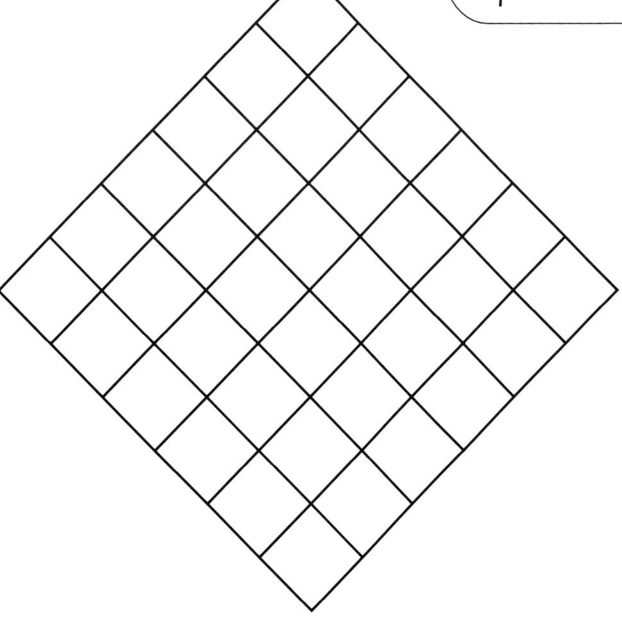

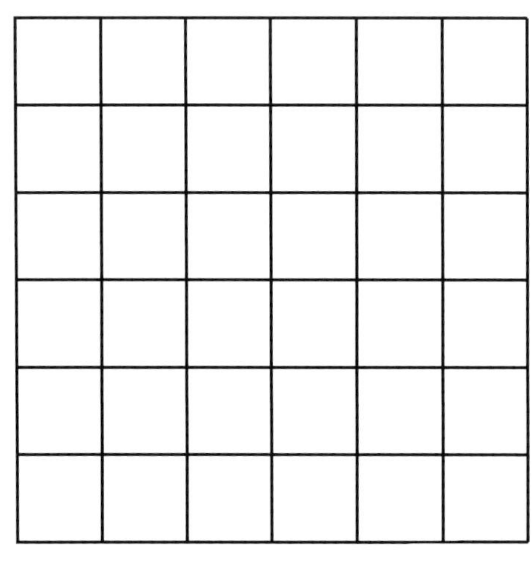

Fill in the missing square numbers.

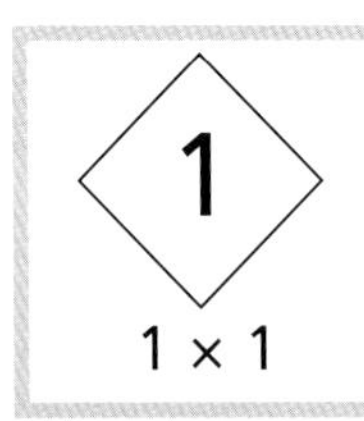
1

1 × 1

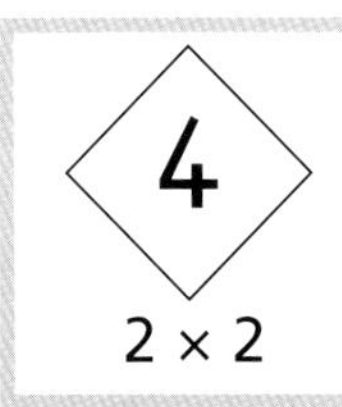
4

2 × 2

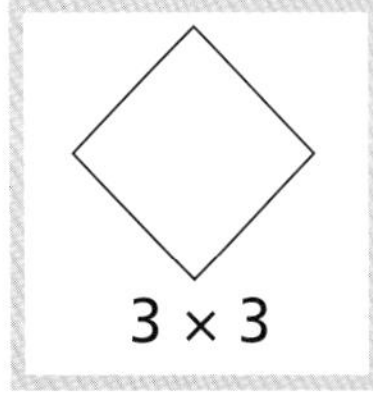

3 × 3

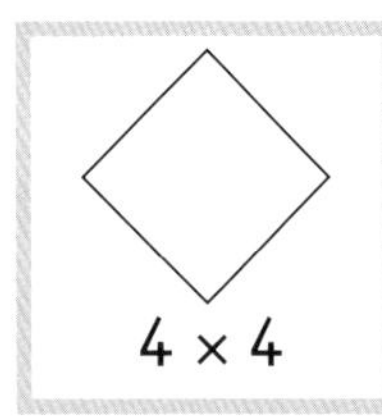

4 × 4

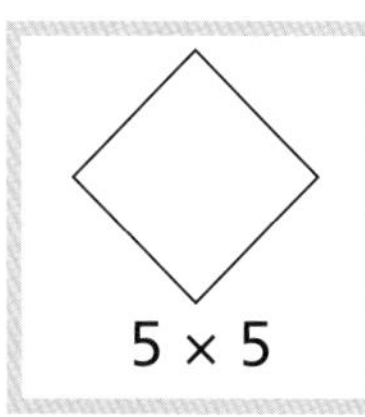

5 × 5

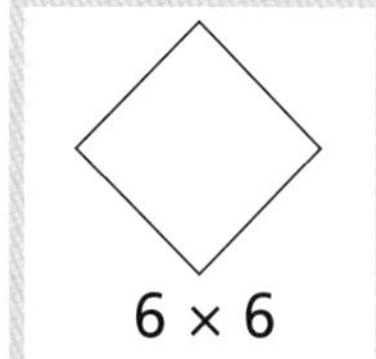

6 × 6

Shopping

Name ____________________

Date ____________________

Use notes and coins.
Add up and write the total.

75p + 75p = ______

95p + £8·50 = ______

£4·40 + £3·60 = ______

£12·15 + £2·35 = ______

£2·99 + £9·99 = ______

£10·00 + £4·99 = ______

£1·97 + £5·60 = ______

£9·45 + £2·95 = ______

Shopping

Y22

Name ____________________

Date ____________________

Make up some questions for a friend.
Make up the prices ...

£3·50
I had £ 10
How much change? ______

and write how much money I had.

These questions are for ______________________________.

£ ____

I had £ ______

How much change? ________

£ ____

I had £ ______

How much change? ________

£ ____

I had £ ______

How much change? ________

£ ____

I had £ ______

How much change? ________

£ ____

I had £ ______

How much change? ________

£ ____

I had £ ______

How much change? ________

Lunch

Name ____________________

Date ____________________

Y23

APPLE AND BLACKCURRANT DRINK
TROPICAL DRINK
ORANGE DRINK

3 drinks in a pack.
Fill in the missing numbers.

Number of packs	0	1	2	3	4	5	6	7	8	9	10	11	12
How many drinks?	0	3											

I need 6 drinks.
How many packs?

$6 \div 3 = 2$
2 packs

21 drinks. How many packs? ________

15 drinks. How many packs? ________

27 drinks. How many packs? ________

24 drinks. How many packs? ________

$3 \times 8 =$ _____ $12 \div 3 =$ _____ $7 \times 3 =$ _____

$3 \times 6 =$ _____ $18 \div 3 =$ _____ $3 \times 3 =$ _____

$3 \times 9 =$ _____ $30 \div 3 =$ _____ $4 \times 3 =$ _____

$$\begin{array}{r} 11 \\ \times\ 3 \\ \hline \\ \hline \end{array} \qquad \begin{array}{r} 12 \\ \times\ 3 \\ \hline \\ \hline \end{array} \qquad \begin{array}{r} 15 \\ \times\ 3 \\ \hline \\ \hline \end{array}$$

Lunch

Name ____________________

Date ____________________

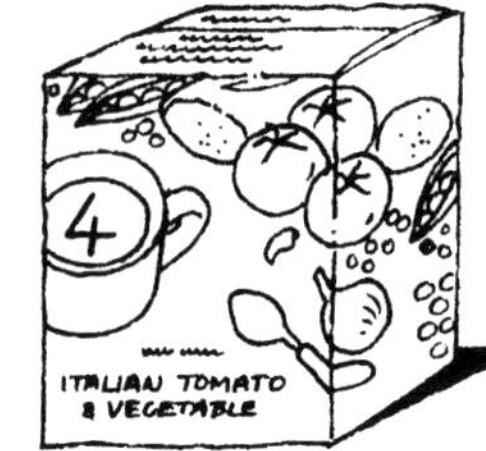
4
ITALIAN TOMATO
& VEGETABLE

4 soups in a pack.
Fill in the missing numbers.

Number of packs	0	1	2	3	4	5	6	7	8	9	10	11	12
How many soups?	0	4	8										

I need 8 soups.
How many packs?

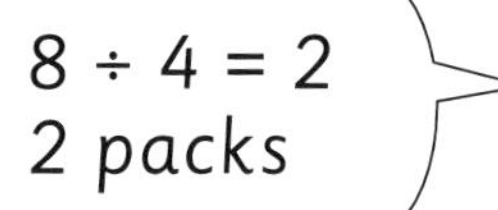
8 ÷ 4 = 2
2 packs

24 soups. How many packs? ________

40 soups. How many packs? ________

28 soups. How many packs? ________

32 soups. How many packs? ________

$4 \times 9 =$ ____ $20 \div 4 =$ ____ $7 \times 4 =$ ____

$4 \times 4 =$ ____ $36 \div 4 =$ ____ $3 \times 4 =$ ____

$4 \times 5 =$ ____ $8 \div 4 =$ ____ $8 \times 4 =$ ____

$$\begin{array}{r} 12 \\ \times\ 4 \\ \hline \\ \hline \end{array} \qquad \begin{array}{r} 13 \\ \times\ 4 \\ \hline \\ \hline \end{array} \qquad \begin{array}{r} 16 \\ \times\ 4 \\ \hline \\ \hline \end{array}$$

Quarter each

Name ______________________

Date ______________________

Y25

You can always share a <u>multiple of 4</u> between four.

A quarter each.

Use coins.

A quarter of 16p is ______

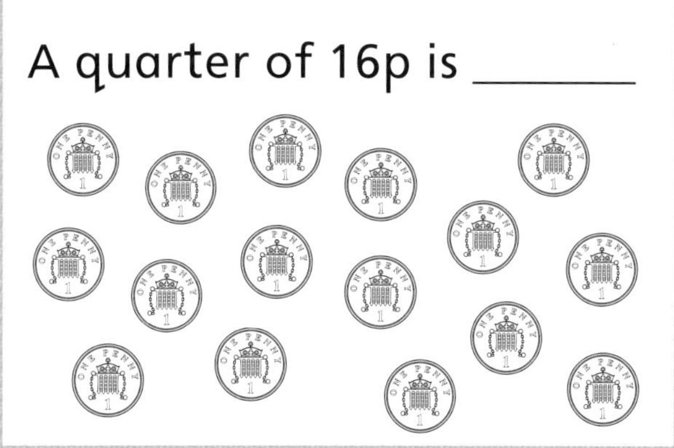

A quarter of 24p is ______

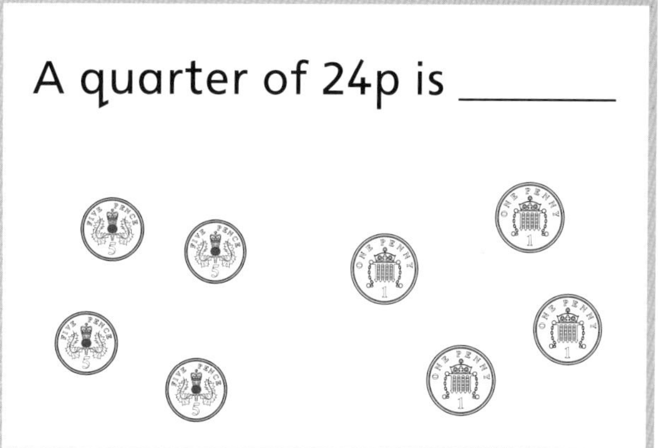

A quarter of 32p is ______

A quarter of 48p is ______

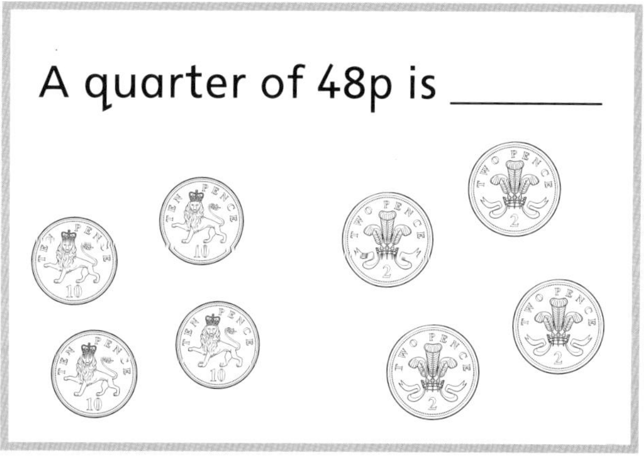

$\frac{1}{4}$ is a short way of writing a quarter.

$\frac{1}{4}$ of 20p is ________ 20 ÷ 4 = ______

$\frac{1}{4}$ of 40p is ________ 40 ÷ 4 = ______

$\frac{1}{4}$ of 60p is ________ 60 ÷ 4 = ______

$\frac{1}{4}$ of 80p is ________ 80 ÷ 4 = ______

◀ Yellow Pupil Book Part 1 pages 34 and 35
▶ Copymaster Y26

Quarter each

Name ____________________

Date ____________________

Use tens and ones.

Swap a ten for ten ones if you need to.

44

$\frac{1}{2}$ of 44 is ________

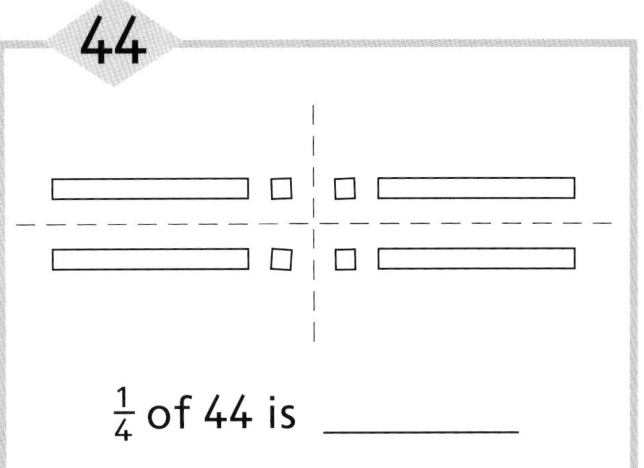

44

$\frac{1}{4}$ of 44 is ________

Find half ...
$\frac{1}{2}$ of 8 is 4
then half again, to get $\frac{1}{4}$
$\frac{1}{4}$ of 8 is 2

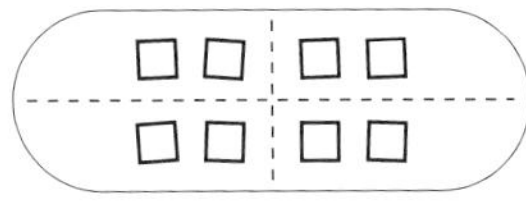

$\frac{1}{2}$ of 20 = ________	$\frac{1}{4}$ of 20 = ________
$\frac{1}{2}$ of 28 = ________	$\frac{1}{4}$ of 28 = ________
$\frac{1}{2}$ of 36 = ________	$\frac{1}{4}$ of 36 = ________
$\frac{1}{2}$ of 44 = ________	$\frac{1}{4}$ of 44 = ________
$\frac{1}{2}$ of 52 = ________	$\frac{1}{4}$ of 52 = ________
$\frac{1}{2}$ of 60 = ________	$\frac{1}{4}$ of 60 = ________
$\frac{1}{2}$ of 68 = ________	$\frac{1}{4}$ of 68 = ________

Y27
T

Times tables bingo

sheet 1 of 2

GP

Print on card if possible. Reusable.
Cut out the instructions card, 4 bingo cards, and 15 question cards.
Store in a clear zip-top wallet or in an envelope. If possible, include 32 counters.

≈ Times tables bingo ≈

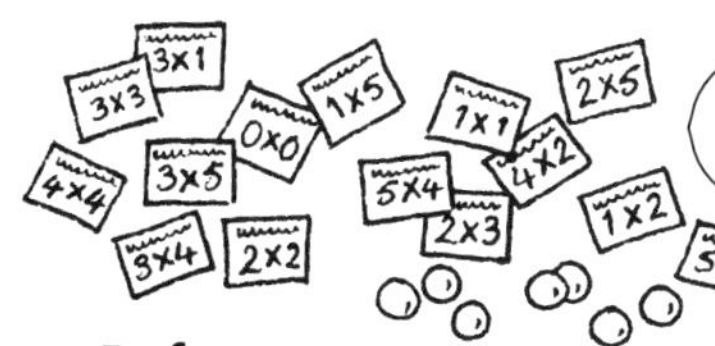
3x1
3x3
1x5
0x0
2x5
1x1
4x4
3x5
5x4
4x2
2x3
1x2
5x5
3x4
2x2

A game for 2, 3 or 4 people.

Times tables bingo
9
0
3
15
25
1
10
6

- **Before you start**
  You need a bingo card and 8 counters each.
  Shuffle the 15 question cards.
  Put them in a pile, face down.

- **How to play**

Take a question card and read it to everyone.

If the answer is on their bingo card, they cover it with a counter.

Now it is your friend's go.

- **Keep going until someone says 'Bingo' because they have covered all their numbers.**

◀ Yellow Pupil Book Part 1; **Mental recall of tables facts within 25**

Number Connections © Rose Griffiths 2005
Harcourt Education Ltd

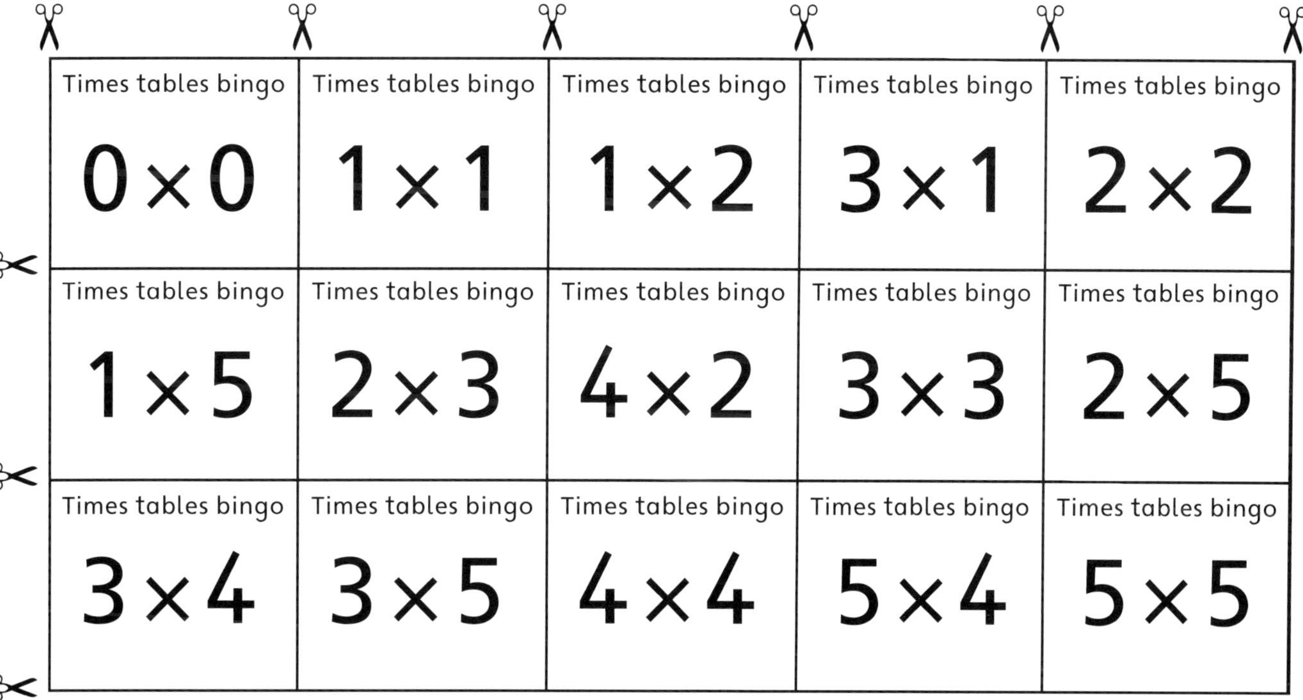

Times tables bingo	Times tables bingo	Times tables bingo	Times tables bingo	Times tables bingo
0×0	1×1	1×2	3×1	2×2
Times tables bingo	Times tables bingo	Times tables bingo	Times tables bingo	Times tables bingo
1×5	2×3	4×2	3×3	2×5
Times tables bingo	Times tables bingo	Times tables bingo	Times tables bingo	Times tables bingo
3×4	3×5	4×4	5×4	5×5

Times tables bingo

sheet 2 of 2

Print on card if possible. Reusable.

Times tables bingo			
3		15	9
	25	2	
12	4		16

Times tables bingo			
6	25	3	
12		16	9
	8		0

Times tables bingo			
	0		9
25		3	15
10	6	1	

Times tables bingo			
4	12	6	
16		20	5
	2		25

Calculator race

Y29
T

sheet 1 of 2

Print on card if possible. Reusable.
Cut out the instructions card and 72 number cards.
Store in a clear zip-top wallet or in a small sandwich box.

≈ Calculator race ≈

A game for 2 people.

- **Before you start**
  You need paper, a pencil and a calculator.
  Put the number cards in a tub or dish.

- **How to play**

Take 2 number cards. Put them on the table.
Add them in your head or on paper ...

while your friend adds them
on the calculator.

The first person to get
the right answer wins the cards.

Now it is your
friend's go.

- **You can keep playing until all the cards have gone.**
  Count how many cards you won.

◀ Yellow Pupil Book Part 1; **Addition within 80**

Calculator race

sheet 2 of 2

Print on card if possible. Reusable. Cut into 72 number cards.

5	6 six	7	8	9 nine	10	11	12	13
14	15	16	17	18	19	20	21	22
23	24	25	26	27	28	29	30	31
32	33	34	35	36	37	38	39	40
1	2	3	4	5	6 six	7	8	9 nine
10	11	12	13	14	15	16	17	18
19	20	21	22	23	24	25	26	27
28	29	30	31	32	10	20	30	40

Y31

One hundred and eighty

Name ______________________

Date ______________________

How many in each box?

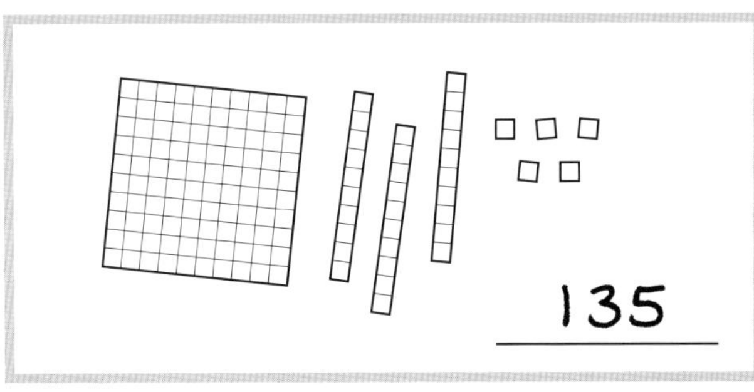
135

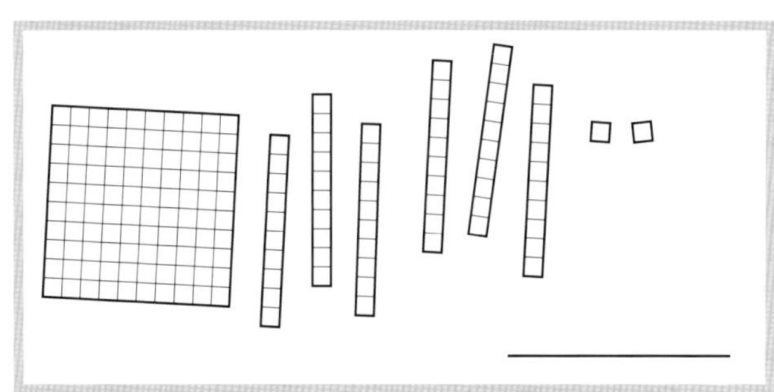

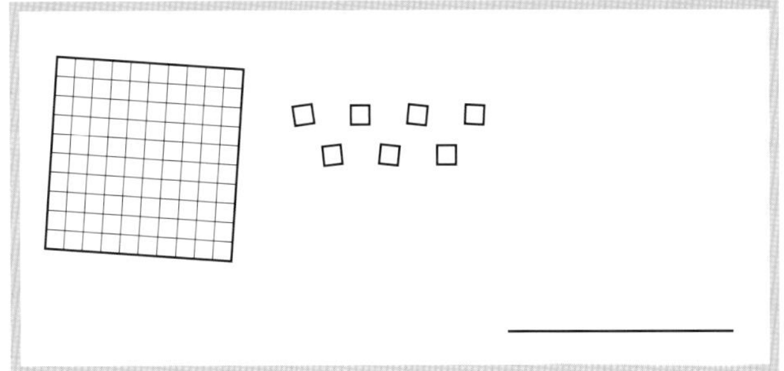

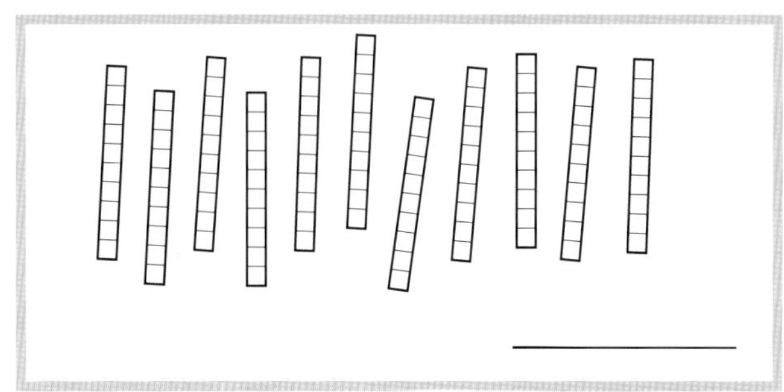

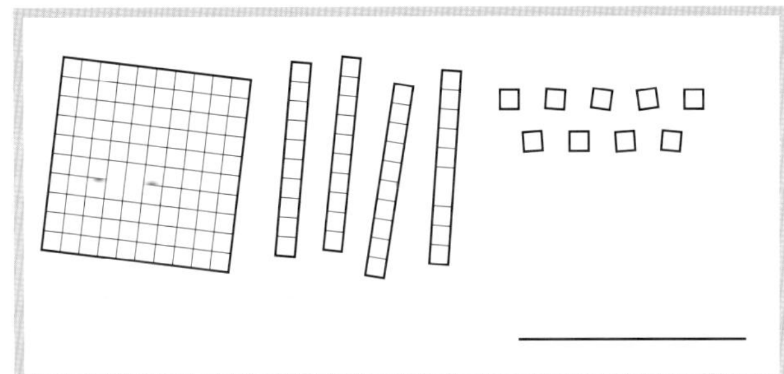

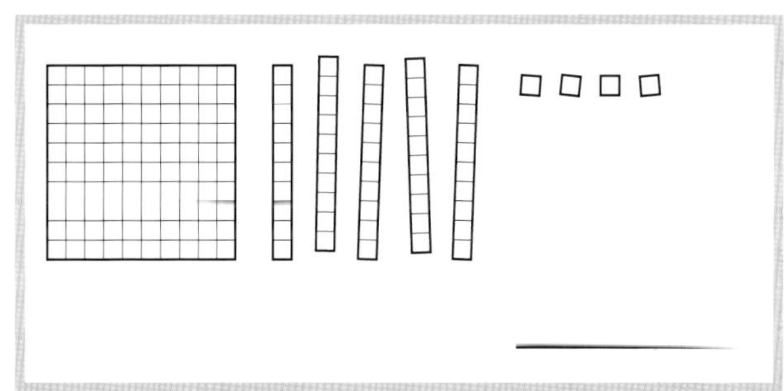

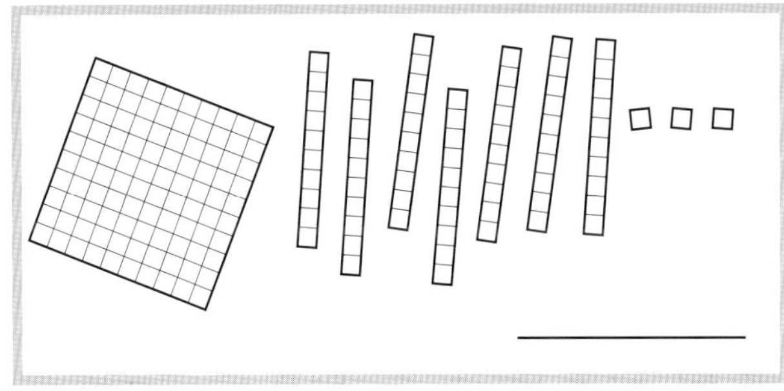

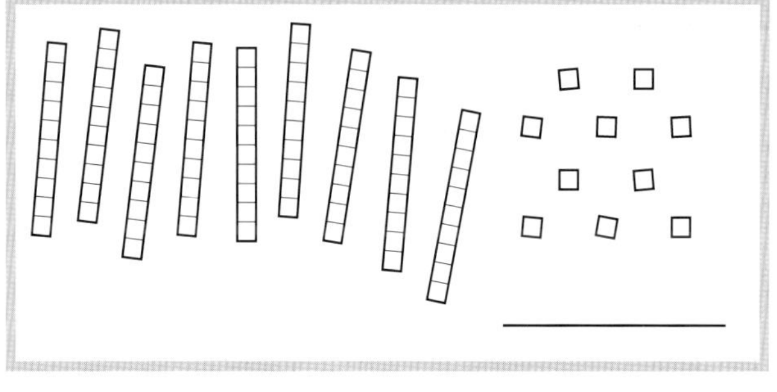

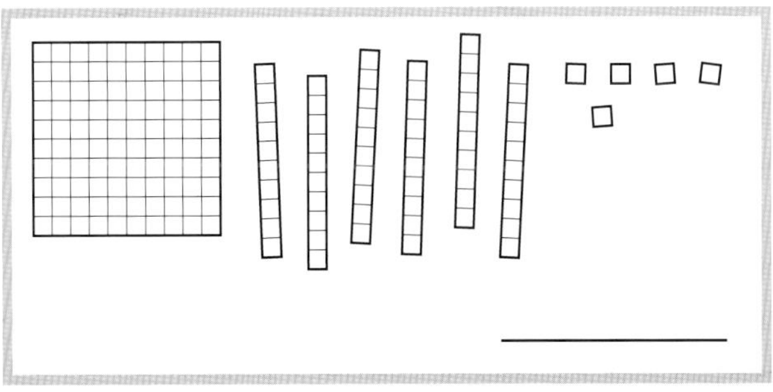

How tall are you?

Name ____________________

Date ____________________

Use with Copymaster Y33.

- Write your name (on Y33) and the missing numbers.
- Cut out the strips.
- Glue them together in the right order.
  Make sure you keep the strips in line.
- Colour the tens.

Make your own height chart.

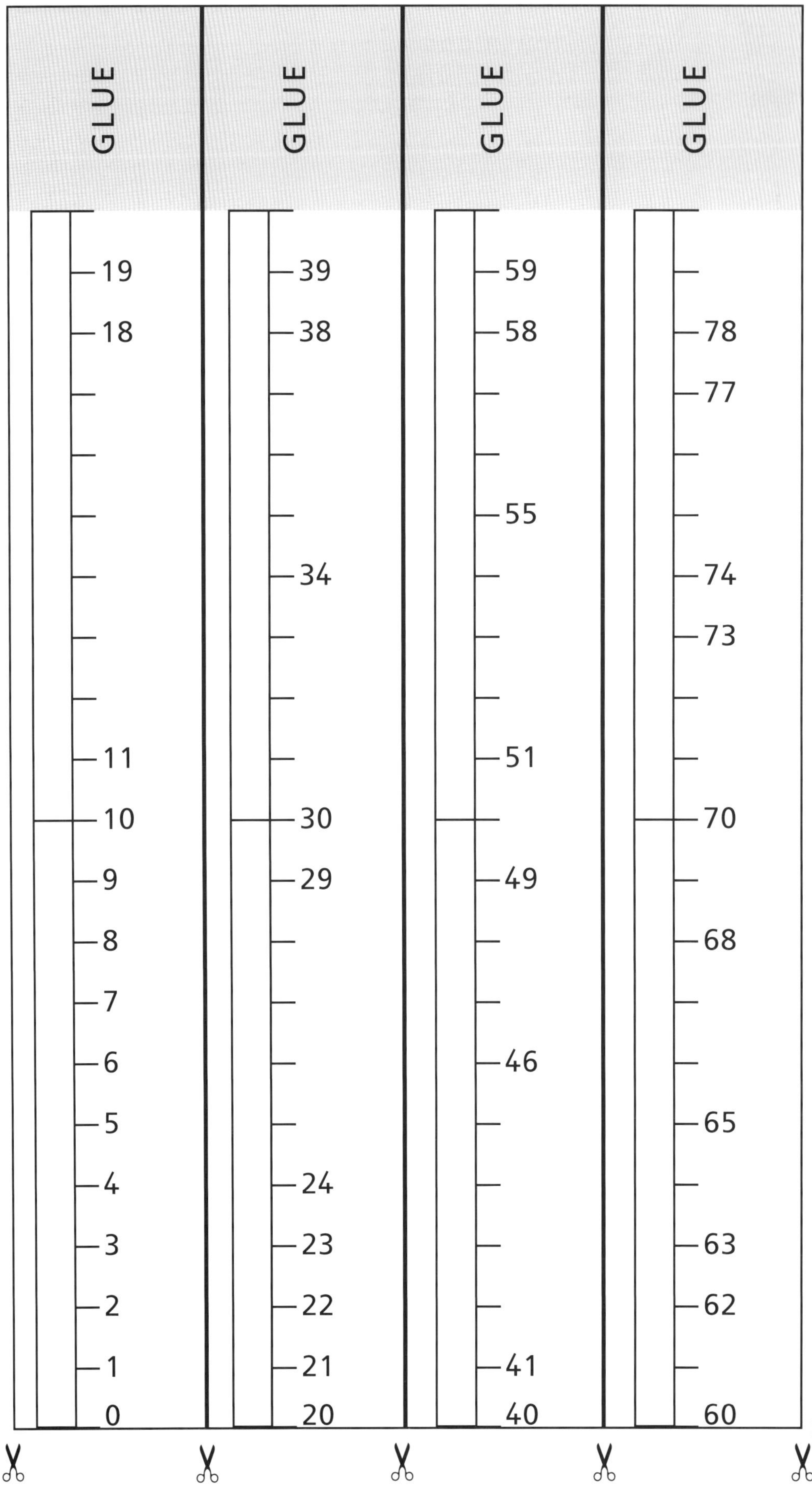

GLUE	GLUE	GLUE	GLUE
19	39	59	
18	38	58	78
			77
		55	
	34		74
			73
11		51	
10	30		70
9	29	49	
8			68
7			
6		46	
5			65
4	24		
3	23		63
2	22		62
1	21	41	
0	20	40	60

How tall are you?

Name ____________________

Date ____________________

Use with Copymaster Y32.

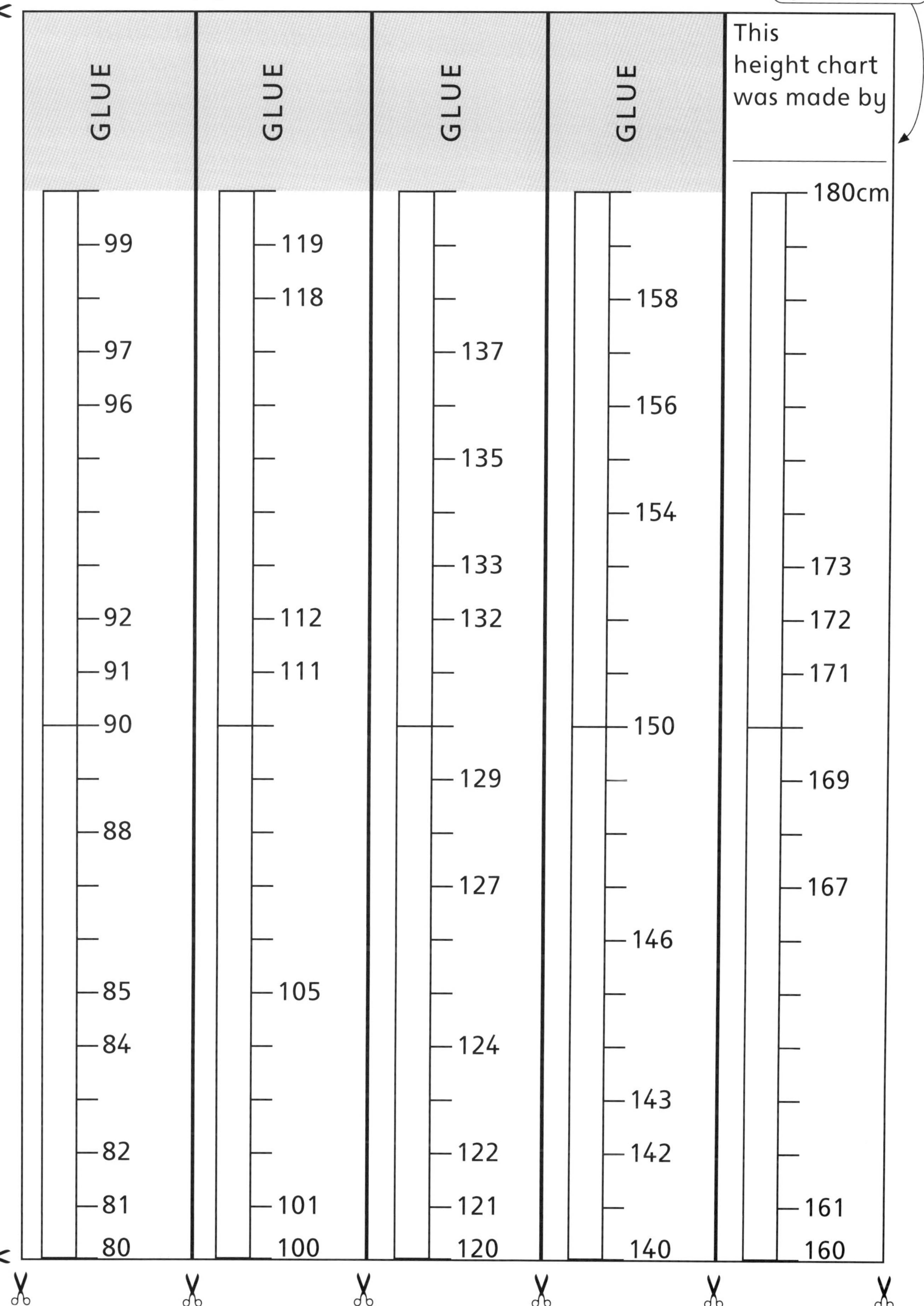

your name
This height chart was made by
GLUE
GLUE
GLUE
GLUE
99
97
96
92
91
90
88
85
84
82
81
80
119
118
112
111
105
101
100
137
135
133
132
129
127
124
122
121
120
158
156
154
150
146
143
142
140
180cm
173
172
171
169
167
161
160

Half price sale

Name ______________________

Date ______________________

How much do these cost in the sale?

Everything
half price

~~£8·50~~

~~£14·00~~

~~£1·00~~

~~£5·00~~

~~£22·00~~

~~£5·50~~

~~£7·50~~

CARTOON
TIME
VIDEO COLLECTION

~~£13·00~~

~~£18·50~~

◀ Yellow Pupil Book Part 2 pages 42 and 43
▶ Copymaster Y35

Half price sale

Y35

Name ______________________

Date ______________________

Make up questions for a friend, like the ones on Copymaster Y34.

These questions are for ______________________ .

Everything
half price

How much do these cost in the sale?

£	£	£
£	£	£

✓ or ✗

Tables stars

Name ______________________

Date ______________________

Use these cards to practise your tables up to 6 × 6.

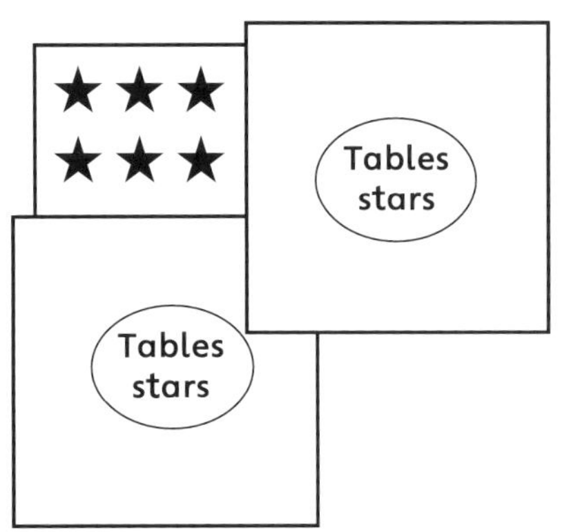
2 × 3 = 6
Tables stars
Tables stars

1 Colour each row of stars in a different colour (like the ones on Pupil Book page 44).

2 Cut out the 3 cards. Write your name on the back of each card.

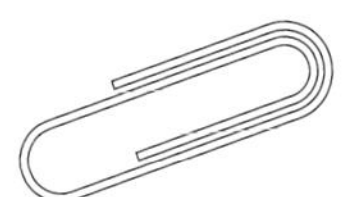

3 Paperclip them together to keep in your book.

Tables stars

Tables stars

☆ ☆ ☆ ☆ ☆ ☆
☆ ☆ ☆ ☆ ☆ ☆
☆ ☆ ☆ ☆ ☆ ☆
☆ ☆ ☆ ☆ ☆ ☆
☆ ☆ ☆ ☆ ☆ ☆
☆ ☆ ☆ ☆ ☆ ☆

Y37

Tables stars

Name ____________________

Date ____________________

2 rows of 3 stars …

2	×	3	=	6

Fill in the missing numbers.

	×		=	

	×		=	

	×		=	

	×		=	

	×		=	

	×		=	

Ice pops

Name ______________________

Date ______________________

Colour each thermometer to show the temperature.

It's hot!
25 degrees.
°C
50
40
30
20
10
0
-10
-20
Do I need my coat?
15 degrees.
°C
50
40
30
20
10
0
-10
-20

Minus
5 degrees.

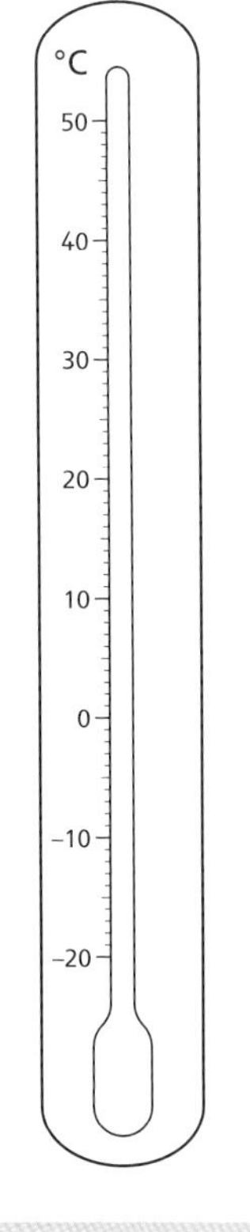
°C
50
40
30
20
10
0
-10
-20

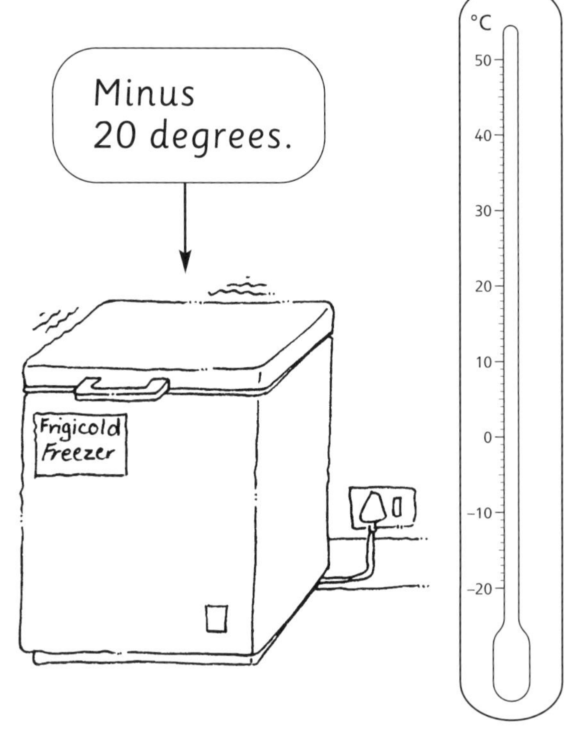
Minus
20 degrees.
Frigicold
Freezer
°C
50
40
30
20
10
0
-10
-20

Ice pops

Name ______________________

Date ______________________

Y39

Fill in the missing numbers.

Write small!

°C

50

45

40

30

20

10

0

–10

–20

12

11

10

9

8

0

–1

–2

–3

–10

Multiples of 5 and 10; negative numbers ◀ Yellow Pupil Book Part 2 pages 46 and 47
▶ Copymaster Y40

Ice pops

Name ____________________

Date ____________________

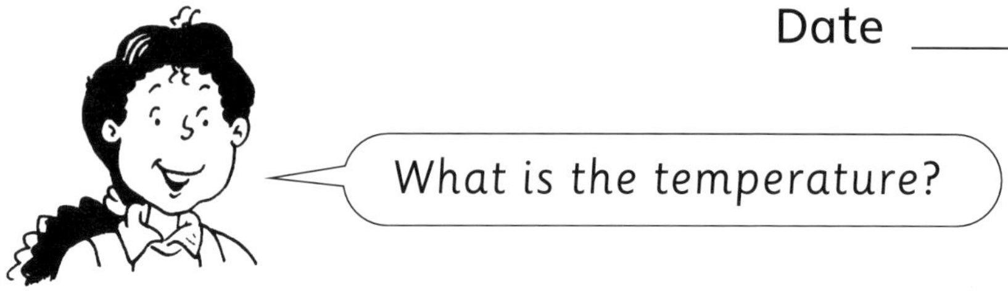

What is the temperature?

°C	°C	°C	°C
50	50	50	50
40	40	40	40
30	30	30	30
20	20	20	20
10	10	10	10
0	0	0	0
–10	–10	–10	–10
–20	–20	–20	–20

°C	°C	°C	°C
50	50	50	50
40	40	40	40
30	30	30	30
20	20	20	20
10	10	10	10
0	0	0	0
–10	–10	–10	–10
–20	–20	–20	–20

Y41

More tables stars

Name ____________________

Date ____________________

How many 2s in 8?
4

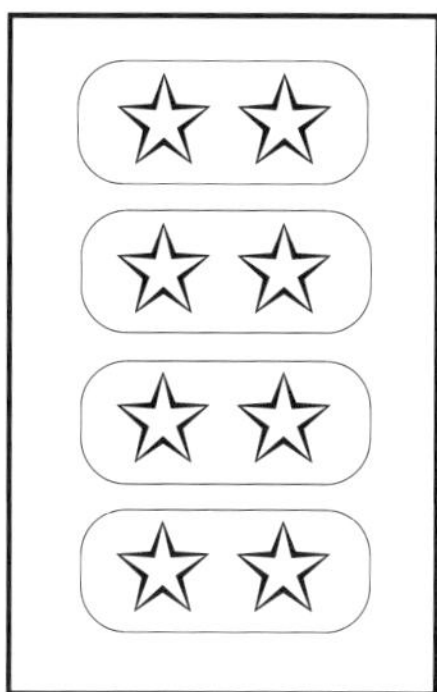

$2 \overline{)8}$ 4

$3 \overline{)15}$	$4 \overline{)16}$	$4 \overline{)8}$
$6 \overline{)30}$	$6 \overline{)18}$	$5 \overline{)25}$
$3 \overline{)12}$	$5 \overline{)20}$	$6 \overline{)6}$

More tables stars

Name ____________________

Date ____________________

Do each division like this …

then check on a calculator.

$3\overline{)18}$ → 6

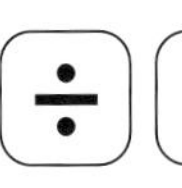

1 8 ÷ 3 =

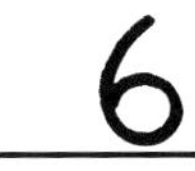

$6\overline{)36}$

3 6 ÷ =

$4\overline{)20}$

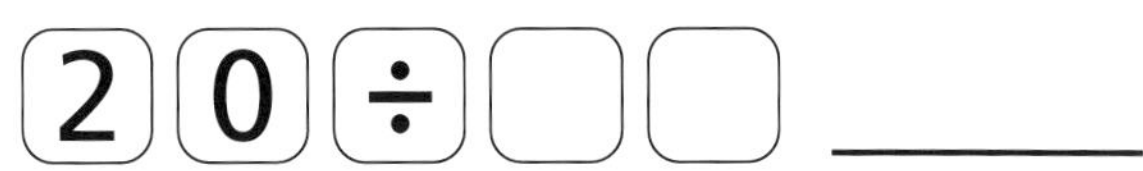
2 0 ÷

$5\overline{)30}$

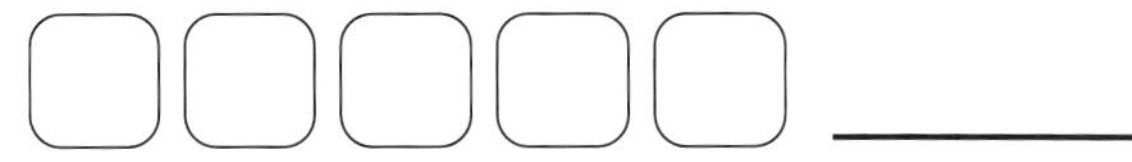

$3\overline{)9}$

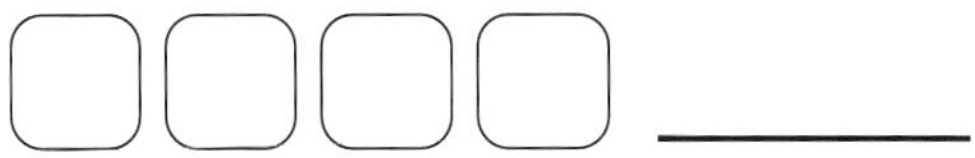

$6\overline{)24}$

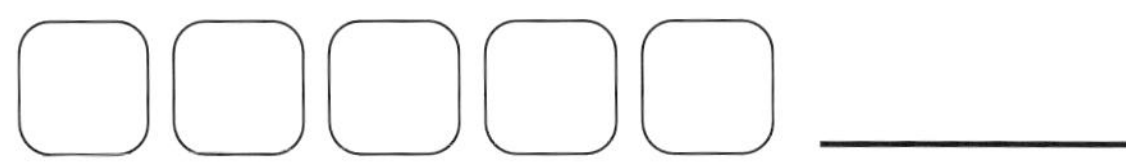

$2\overline{)10}$

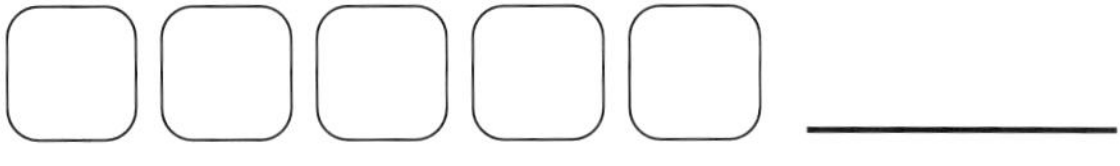

$4\overline{)24}$

Y43

Speedy tables C

1 2 3 minute test

Name ______________________

Date ______________________

3 × 6 = ____	100 ÷ 10 = ____	4 × 7 = ____
10 × 0 = ____	27 ÷ 3 = ____	16 ÷ 2 = ____
2 × 9 = ____	12 ÷ 4 = ____	5 × 8 = ____
4 × 8 = ____	35 ÷ 5 = ____	30 ÷ 5 = ____
7 × 3 = ____	24 ÷ 3 = ____	5 × 5 = ____
4 × 4 = ____	60 ÷ 10 = ____	24 ÷ 4 = ____
5 × 9 = ____	36 ÷ 4 = ____	 **Score:** ____

Mental recall of tables facts ◀ Yellow Pupil Book Part 2; pages 50 and 51 onwards

Speedy tables  D

1 2 3 minute test

Name ______________________

Date ______________________

3 × 3 = ____	28 ÷ 4 = ____	10 × 10 = ____
5 × 7 = ____	18 ÷ 6 = ____	32 ÷ 4 = ____
3 × 9 = ____	40 ÷ 5 = ____	0 × 9 = ____
1 × 8 = ____	80 ÷ 10 = ____	18 ÷ 2 = ____
9 × 4 = ____	7 ÷ 7 = ____	4 × 6 = ____
2 × 7 = ____	45 ÷ 5 = ____	30 ÷ 5 = ____
8 × 3 = ____	8 ÷ 2 = ____	 **Score:** ____

Keeping fit

Name ____________________

Date ____________________

I went swimming every day on holiday.

How many lengths altogether, in the 1st week?

____

1st week	Lengths I swam each day
Mon. 7	Fri. 7
Tues. 3	Sat. 8
Wed. 5	Sun. 6
Thurs. 7	

How many lengths altogether, in the 2nd week?

____

2nd week	Lengths I swam each day
Mon. 7	Fri. 4
Tues. 7	Sat. 5
Wed. 3	Sun. 9
Thurs. 5	

How many lengths altogether, in 2 weeks? __________

Do these in your head.

44 + 20 = ____

79 + 10 = ____

57 + 30 = ____

62 + 30 = ____

20 + 55 = ____

40 + 23 = ____

✓ or ✗

Keeping fit

Name ______________________

Date ______________________

Use tens and ones if you want to.
Then add on paper.

46 + 37	19 + 58	28 + 71	43 + 37
52 + 18	34 + 52	65 + 23	38 + 29

Use tens and ones.

100 − 62	100 − 47	100 − 98	100 − 54
100 − 33	100 − 71	100 − 15	100 − 26

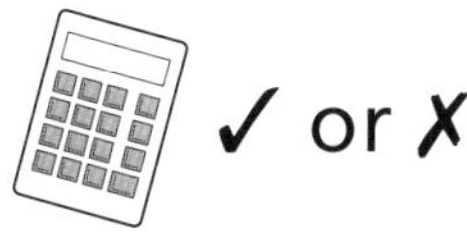

✓ or ✗

Y46

T

Six times table

Name ____________________

Date ____________________

Cut out the eleven tables facts.
Fold along the dotted line and glue flat.

Ask your teacher how to practise with these.

		5×6	30
0×6	0	6×6	36
1×6	6	7×6	42
2×6	12	8×6	48
3×6	18	9×6	54
4×6	24	10×6	60

Six times table

Name ______________________

Date ______________________

Y47

Fill in the missing numbers.
Check with a calculator.

$3 \times 6 = \square$

$18 \div 6 = \square$

$5 \times 6 = \square$

$30 \div 6 = \square$

$10 \times 6 = \square$

$60 \div 6 = \square$

$8 \times 6 = \square$

$48 \div 6 = \square$

$10 \times 6 = \square$

$9 \times 6 = \square$

$\square \times 6 = 48$

$7 \times 6 = \square$

$\square \times 6 = 36$

$5 \times 6 = \square$

$\square \times 6 = 24$

$3 \times 6 = \square$

$2 \times 6 = \square$

$\square \times 6 = 6$

$0 \times 6 = \square$

What is
6 times 7?

____

What is
7 times 6?

____

Multiplying

Name ______________________

Date ______________________

Use tens and ones. Draw them. Then multiply on paper.

19 times 2

19 times 3

19 times 4

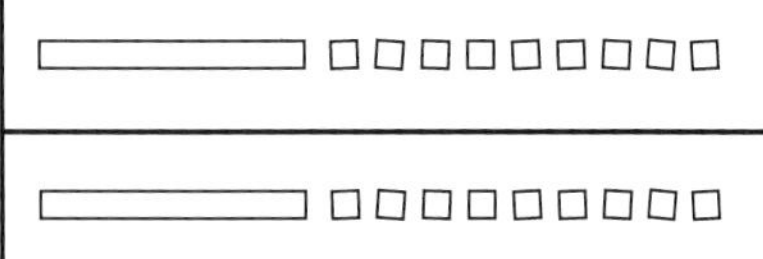

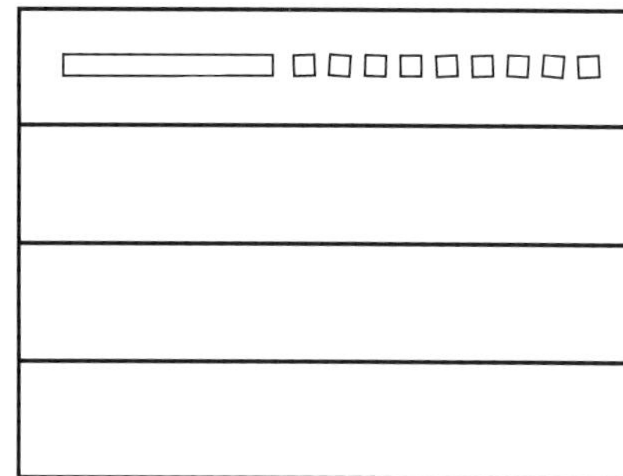

$$\begin{array}{r} 19 \\ \times\ 2 \\ \hline \end{array}$$

$$\begin{array}{r} 19 \\ \times\ 3 \\ \hline \end{array}$$

$$\begin{array}{r} 19 \\ \times\ 4 \\ \hline \end{array}$$

Use tens and ones. Then multiply on paper.

$$\begin{array}{r} 22 \\ \times\ 3 \\ \hline \end{array}$$

$$\begin{array}{r} 18 \\ \times\ 5 \\ \hline \end{array}$$

$$\begin{array}{r} 46 \\ \times\ 2 \\ \hline \end{array}$$

$$\begin{array}{r} 17 \\ \times\ 4 \\ \hline \end{array}$$

$$\begin{array}{r} 16 \\ \times\ 5 \\ \hline \end{array}$$

$$\begin{array}{r} 29 \\ \times\ 3 \\ \hline \end{array}$$

$$\begin{array}{r} 50 \\ \times\ 2 \\ \hline \end{array}$$

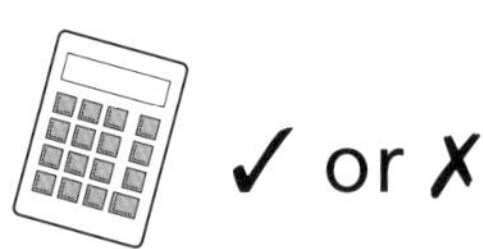
✓ or ✗

◄ Yellow Pupil Book Part 2 pages 56 and 57
► Copymaster Y49

Multiplying

Name ______________________

Date ______________________

Y49

Use tens and ones. Draw them.
Then multiply on paper.

24 times 2

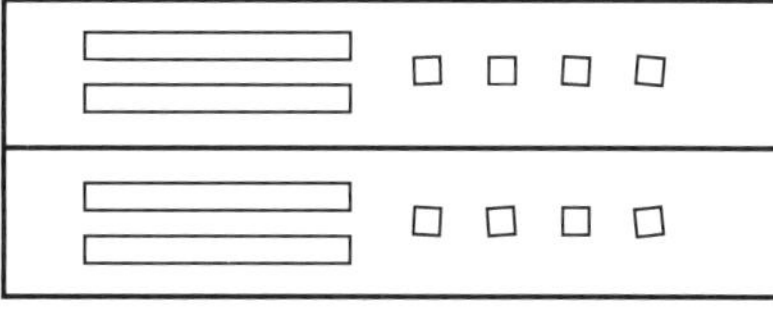

24 times 3

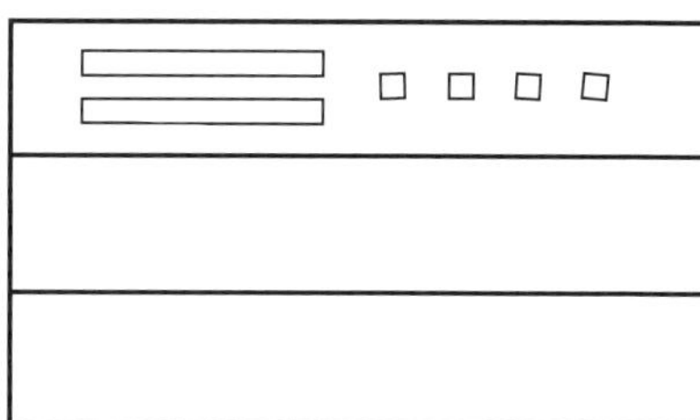

24 times 4

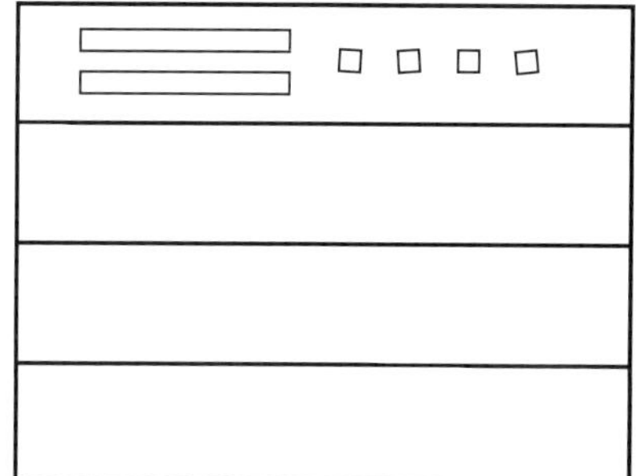

$$\begin{array}{r} 24 \\ \times \quad 2 \\ \hline \\ \hline \end{array} \qquad \begin{array}{r} 24 \\ \times \quad 3 \\ \hline \\ \hline \end{array} \qquad \begin{array}{r} 24 \\ \times \quad 4 \\ \hline \\ \hline \end{array}$$

Multiply on paper. Then use tens and ones.

$$\begin{array}{r} 36 \\ \times \quad 2 \\ \hline \\ \hline \end{array} \qquad \begin{array}{r} 18 \\ \times \quad 3 \\ \hline \\ \hline \end{array} \qquad \begin{array}{r} 41 \\ \times \quad 2 \\ \hline \\ \hline \end{array} \qquad \begin{array}{r} 15 \\ \times \quad 5 \\ \hline \\ \hline \end{array}$$

$$\begin{array}{r} 21 \\ \times \quad 3 \\ \hline \\ \hline \end{array} \qquad \begin{array}{r} 48 \\ \times \quad 2 \\ \hline \\ \hline \end{array} \qquad \begin{array}{r} 23 \\ \times \quad 4 \\ \hline \\ \hline \end{array}$$

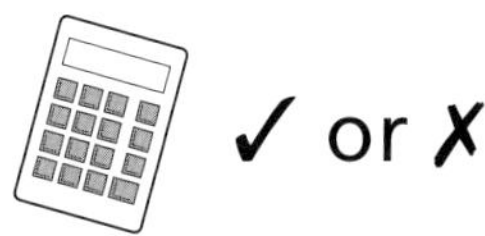
✓ or ✗

What's missing?

Name ______________________

Date ______________________

Write the missing numbers.

1	□	+	7	=	2	2

□	+	6	=	1	5

2	3	–	□	=	2	0

7	□	–	8	=	7	0

□	2	–	5	0	=	2

3	4	+	□	=	4	0

✓ or ✗

Write + or – .

9	□	8	=	1	7

2	7	□	3	=	2	4

4	6	□	7	=	3	9

6	2	□	6	0	=	2

3	5	□	9	=	4	4

1	8	□	1	0	=	8

7	2	□	2	=	7	0

1	5	□	9	=	2	4

What's missing?

Name ____________________

Date ____________________

Y51

Write the missing numbers.

5	×		=	2	0

1	0	÷	2	=	

1	6	÷		=	4

7	×		=	1	4

4	×	8	=	3	

5	×	9	=	4	

3	0	÷		=	6

1		×	8	=	8	0

6	×	1		=	6	0

2	1	÷	3	=	

✓ or ✗

Write × or ÷ .

2	0		4	=	5

1	8		3	=	6

1	0		2	=	2	0

9		4	=	3	6

7		5	=	3	5

2	4		3	=	8

Y52

Dividing marbles

Name ____________________

Date ____________________

Fill in the missing numbers.

$34 \div 2 =$ ______

$2\overline{)34}$

$35 \div 2 =$ ______

$2\overline{)35}$

$29 \div 2 =$ ______

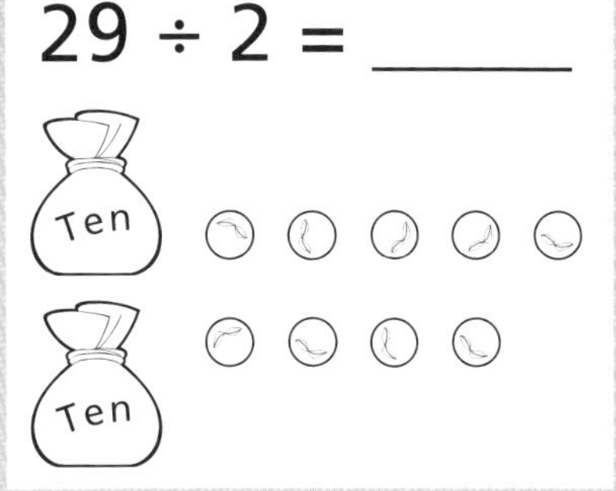
Ten
Ten

$2\overline{)29}$

$31 \div 2 =$ ______

$2\overline{)31}$

$19 \div 2 =$ ______

$2\overline{)19}$

$42 \div 2 =$ ______

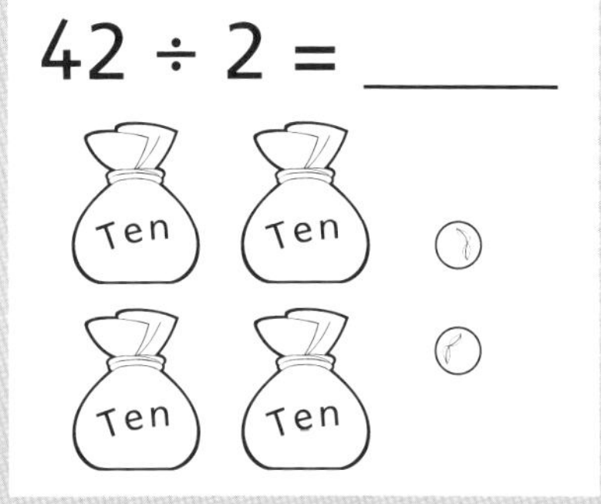
Ten
Ten
Ten
Ten

$2\overline{)42}$

$33 \div 2 =$ ______

$2\overline{)33}$

$50 \div 2 =$ ______

$2\overline{)50}$

$46 \div 2 =$ ______

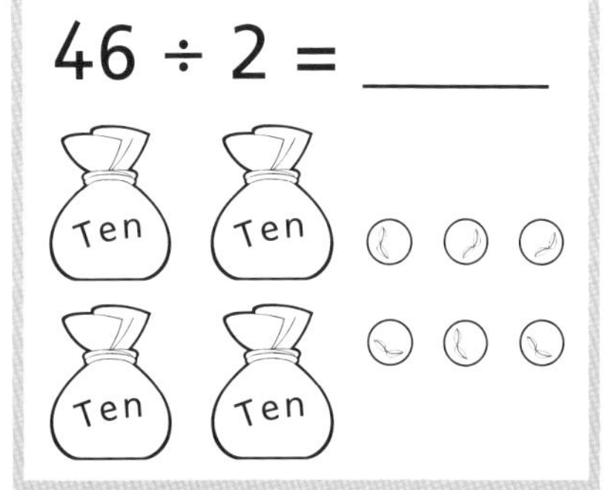
Ten
Ten
Ten
Ten

$2\overline{)46}$

Y53

Dividing marbles

Name ______________________

Date ______________________

Fill in the missing numbers.

64 ÷ 2 = _____

Ten Ten Ten
Ten Ten Ten

2) 64

67 ÷ 2 = _____

Ten Ten Ten
Ten Ten Ten

2) 67

72 ÷ 2 = _____

Ten Ten Ten Ten
Ten Ten Ten

2) 72

56 ÷ 2 = _____

Ten Ten Ten
Ten Ten

2) 56

43 ÷ 2 = _____

Ten Ten
Ten Ten

2) 43

52 ÷ 2 = _____

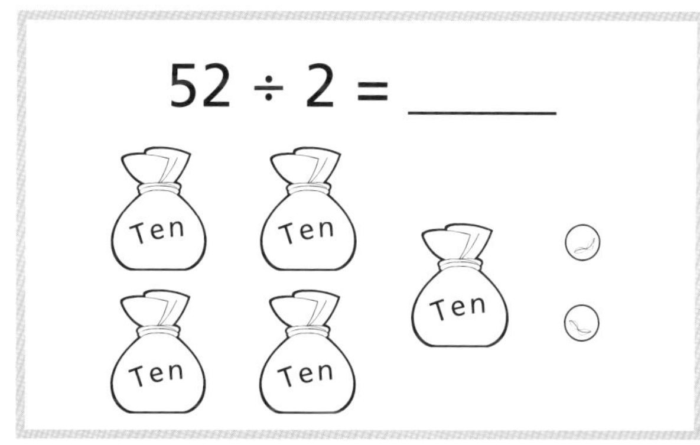
Ten
Ten
Ten
Ten
Ten

2) 52

More dividing

Name ____________________

Date ____________________

Use tens and ones to divide by 2.

How many each? ______

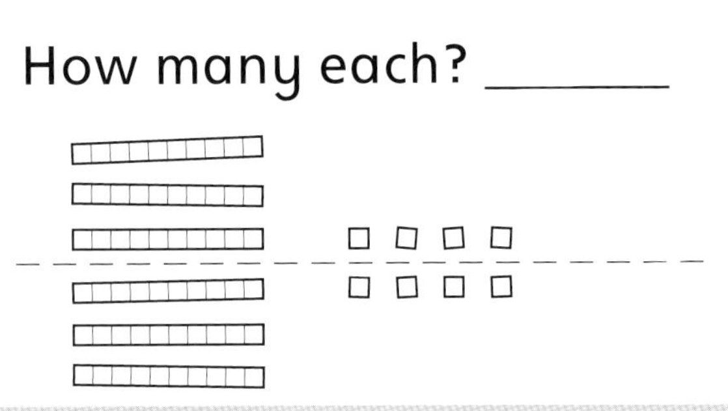

6 8 ÷ 2 = ______

2) 68

How many each? ______

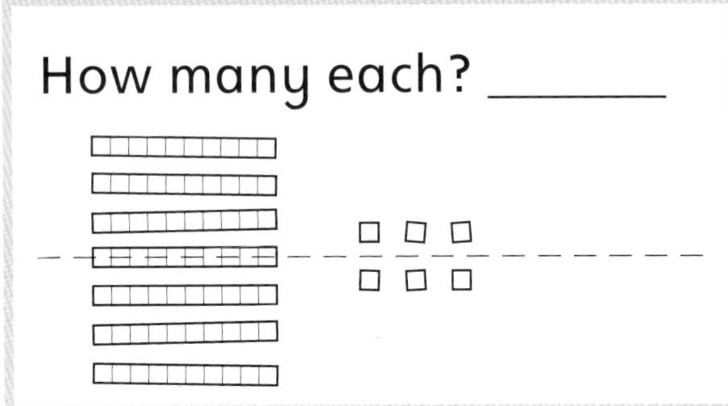

7 6 ÷ 2 = ______

2) 76

How many each? ______

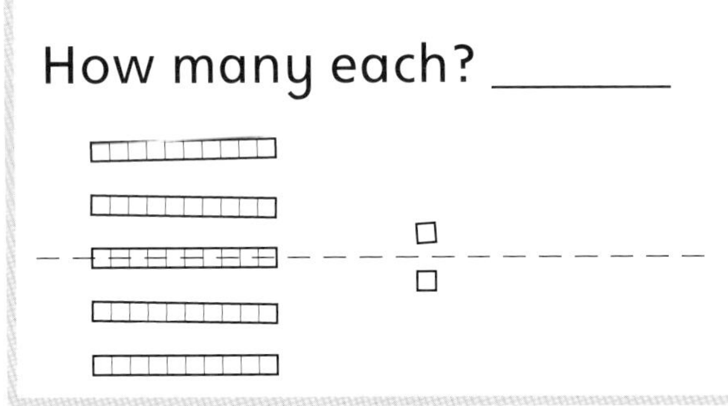

5 2 ÷ 2 = ______

2) 52

Divide on paper. Check with tens and ones.

2) 65   2) 41   2) 78

2) 27   2) 86   2) 94

More dividing

Y55

T

Name ____________________

Date ____________________

Use tens and ones to divide by 2.

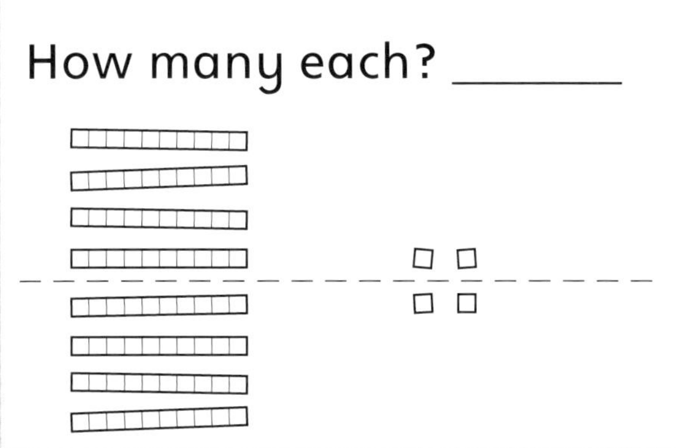
How many each? ______

8 4 ÷ 2 = ______

2) 84

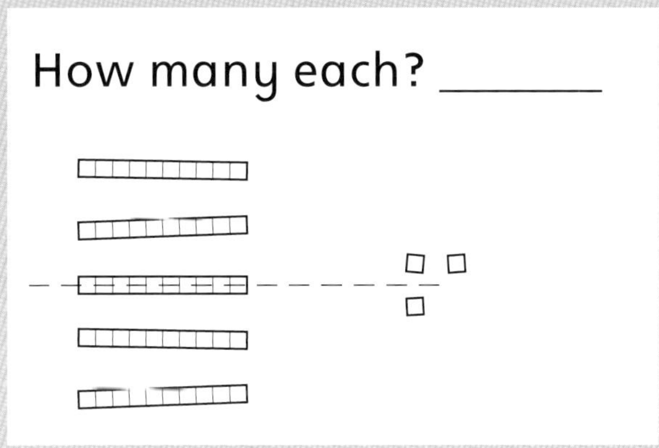
How many each? ______

5 3 ÷ 2 = ______

2) 53

Talk to your teacher about this.

Divide on paper. Check with tens and ones.

2) 52	2) 70	2) 96
2) 85	2) 97	2) 61
2) 73	2) 80	2) 66

Nines and tens

Name ____________________

Date ____________________

Fill in the missing numbers.

29 + 9 = ____
38 + 9 = ____
47 + 9 = ____
56 + 9 = ____
65 + 9 = ____
74 + 9 = ____
83 + 9 = ____

14 + 9 = ____
23 + 9 = ____
32 + 9 = ____
41 + 9 = ____
50 + 9 = ____
59 + 9 = ____
68 + 9 = ____

32 + 20 = ____
32 + 19 = ____

79 + 20 = ____
79 + 19 = ____

56 + 20 = ____
56 + 19 = ____

25 + 20 = ____
25 + 19 = ____

48 + 20 = ____
48 + 19 = ____

64 + 20 = ____
64 + 19 = ____

◀ Yellow Pupil Book Part 2 pages 64 and 65
▶ Copymaster Y57

Nines and tens

Name ____________________

Date ____________________

Y57

Fill in the missing numbers.

$27 + \square = 47$

$27 + 19 = \_\_\_\_$

$61 + \square = 91$

$61 + 29 = \_\_\_\_$

$48 + \square = 58$

$48 + 9 = \_\_\_\_$

$52 + \square = 72$

$52 + 19 = \_\_\_\_$

$46 + \square = 66$

$46 + \square = 56$

$46 + \square = 76$

$46 + \square = 86$

$33 + \square = 43$

$33 + \square = 83$

$33 + \square = 53$

$33 + \square = 63$

$25 + \square = 55$

$25 + \square = 35$

$25 + \square = 65$

$25 + \square = 45$

$49 + \square = 59$

$49 + \square = 79$

$49 + \square = 69$

$49 + \square = 99$

Y58
T

Half price sale

sheet 1 of 2

GP

Print on card if possible. Reusable.
Colour and cut out the instructions card and 20 playing cards.
Store in a clear zip-top wallet or in an envelope. Include token money if possible.

≈ Half price sale ≈

A game for 1, 2 or 3 people.

- **Before you start**
  Shuffle the cards.
  Spread them out on the table, face down.
  You need £20 each (a £10 note, nine £1 coins and two 50p coins) and a money pot.

- **How to play**

Turn over a card.
Say how much the item would cost, half price.

Do you want to buy it?
If you do, put the money in the pot.
If not, turn the card back over.

Now it is your friend's go.

- **Keep going until you can't spend any more money.**

◀ Yellow Pupil Book Part 2; **Halving and doubling money**

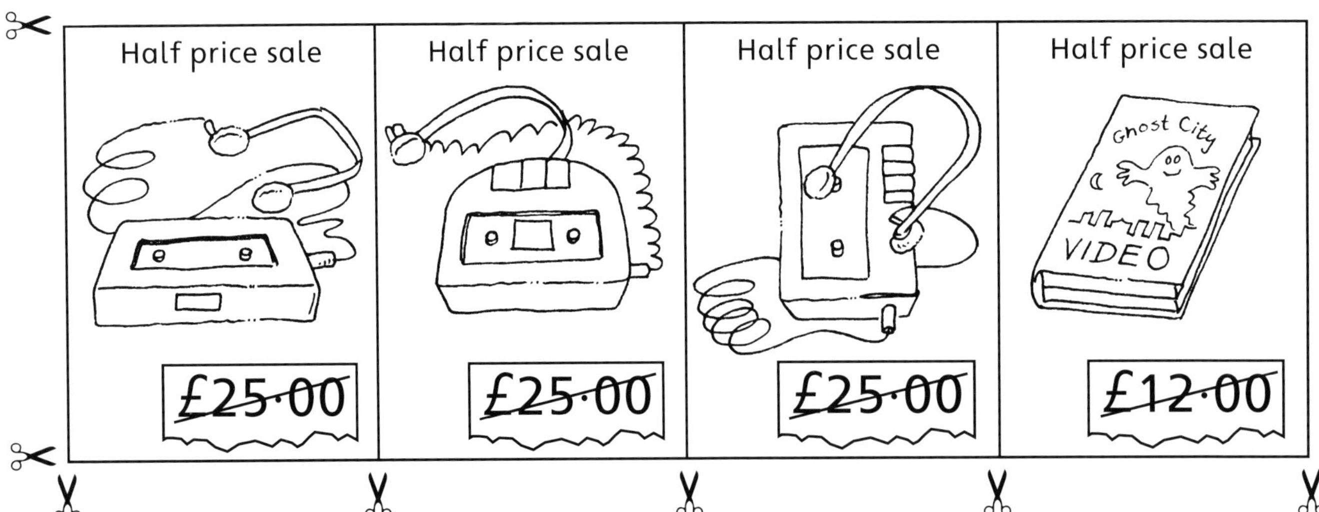
Half price sale
£25·00
Half price sale
£25·00
Half price sale
£25·00
Half price sale
Ghost City
VIDEO
£12·00

◀ Yellow Pupil Book Part 2; use from pages 42 and 43 onwards

Y59
T

Half price sale

sheet 2 of 2

Print on card if possible. Reusable.

GP

Half price sale SPIKE VIDEO ~~£12·00~~	Half price sale SHIPWRECK VIDEO ~~£12·00~~	Half price sale GHOSTS ~~£9·00~~	Half price sale COMPUTERS ~~£9·00~~
Half price sale SOUPS ~~£9·00~~	Half price sale ~~£5·00~~	Half price sale ~~£5·00~~	Half price sale ~~£5·00~~
Half price sale ~~£3·00~~	Half price sale ~~£3·00~~	Half price sale ~~£3·00~~	Half price sale ~~£3·00~~
Half price sale ~~£1·00~~	Half price sale ~~£1·00~~	Half price sale ~~£1·00~~	Half price sale ~~£1·00~~

Y60

T

GP

What's missing?

sheet 1 of 3

Print on card if possible. Reusable.
Cut out the instructions card, 2 sets of number and operations cards, and 2 stands.
Store in a clear zip-top wallet or a storage box.

What's missing?

A game for 2 people.

- **Before you start**
  You each need a calculator, a stand and a set of cards.

- **How to play**

Make up a sum.
You can use 5, 6 or 7 cards.
2 × 3 = 6
Check with a calculator.
Take one card away.
2 3 = 6

Your friend has to work out what's missing.

- **Take turns to make up puzzles.**

◀ Yellow Pupil Book Part 2; **Missing numbers and operations**

What's missing?

sheet 2 of 3

Print 2 copies, on card if possible. Reusable.
(Use 2 different colours of card, to make it easy to sort into 2 sets once cut up.)
Cut into 25 cards.

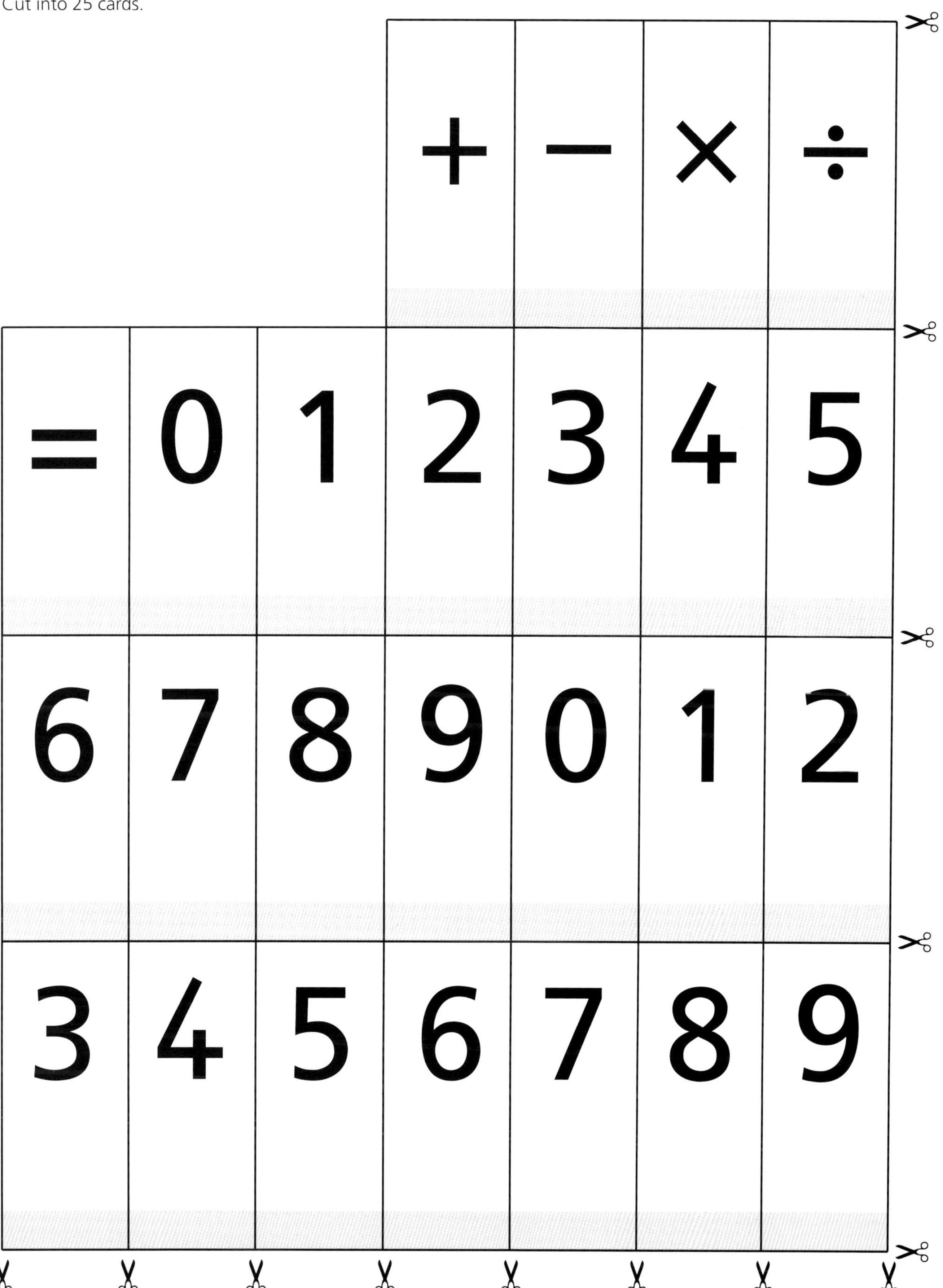
+
−
×
÷
=
0
1
2
3
4
5
6
7
8
9
0
1
2
3
4
5
6
7
8
9

What's missing?

sheet 3 of 3

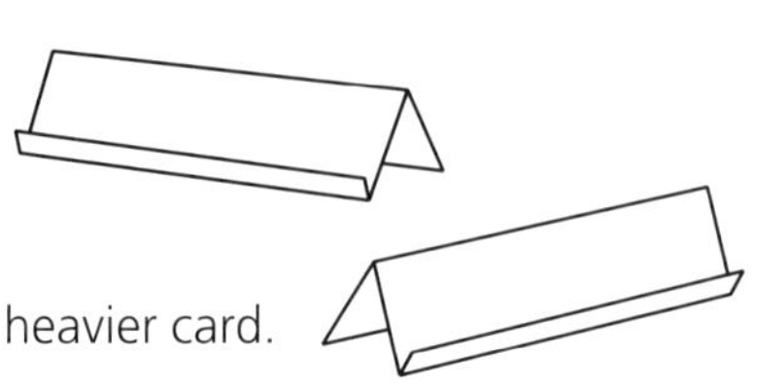

Print on card if possible. Reusable.
Cut out, score and fold along dotted lines.
Alternatively, use these as templates to make your own stands from heavier card.

fold

fold

fold

fold

Two hundred

Name ______________________

Date ______________________

Y63

Make each number with hundreds, tens and ones.

Then draw it.

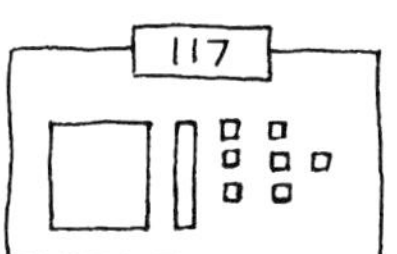
117

123

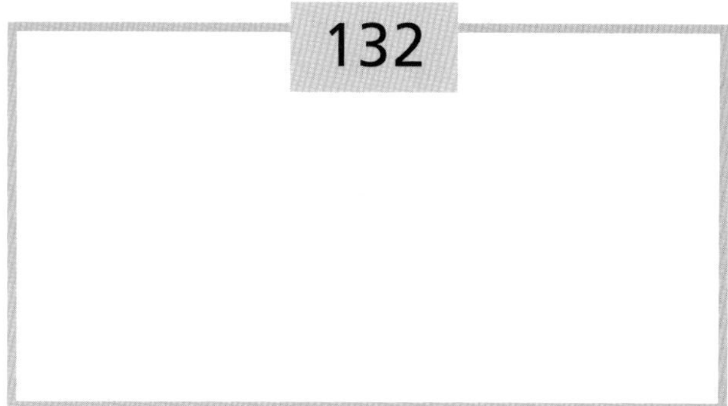

132

194

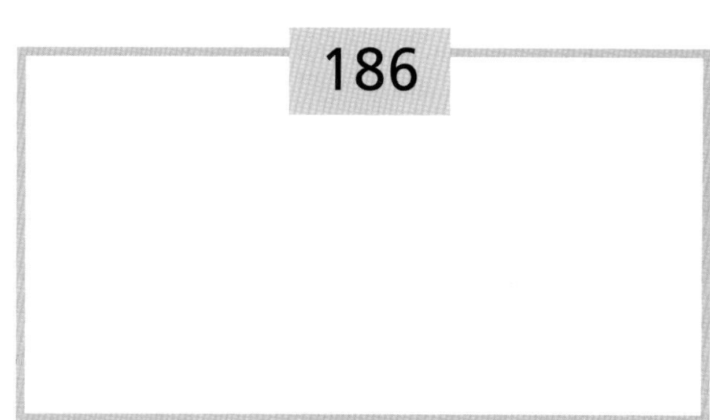

186

106

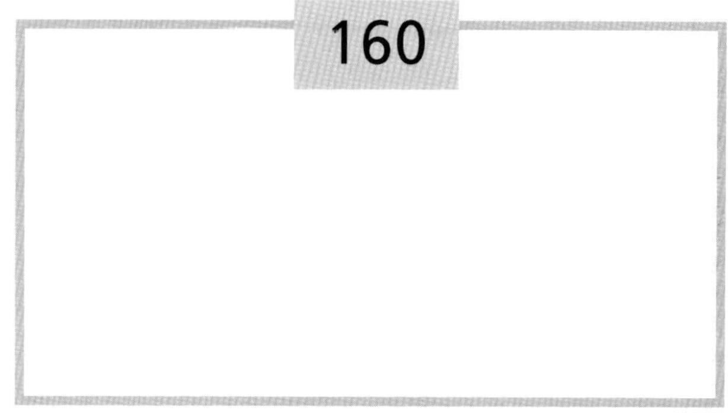

160

199

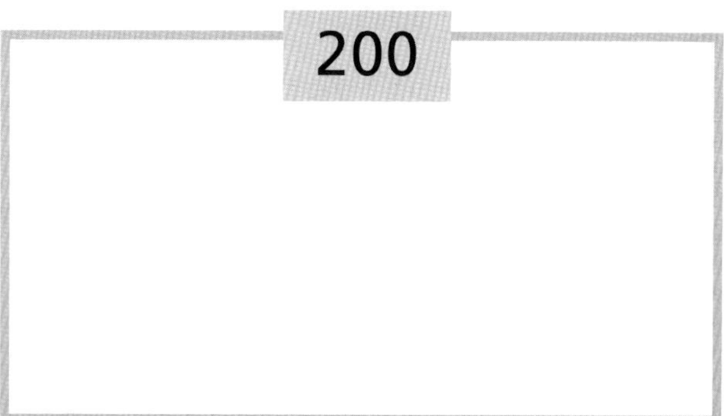

200

Saving

Name ______________________

Date ______________________

How much did we save?

40p a week
for 8 weeks.

____

£2 a month
for 6 months.

____

£2·50 a month
for 3 months.

____

75p a week
for 8 weeks.

____

£1·25 a week
for 6 weeks.

____

£3·50 a month
for 5 months.

____

£5 a month
for 12 months.

____

90p a week
for 5 weeks.

____

◄ Yellow Pupil Book Part 3 pages 70 and 71
► Copymaster Y65

Saving

Name ____________________

Date ____________________

Y65

Make up questions for a friend.

When they finish, ✓ or ✗.

These questions are for ____________________ .

How much did we save?

______ a month for _____ months.

______

______ a week for _____ weeks.

______

______ a week for _____ weeks.

______

______ a month for _____ months.

______

______ a month for _____ months.

______

______ a week for _____ weeks.

______

Teen sums

Name ____________________

Date ____________________

Fill in the missing numbers.

1 + ☐ = 16

6 + ☐ = 16

9 + ☐ = 16

2 + ☐ = 16

11 + ☐ = 16

7 + ☐ = 16

4 + ☐ = 16

16 + ☐ = 16

14 + ☐ = 16

☐ + 0 = 16

☐ + 5 = 16

☐ + 12 = 16

☐ + 3 = 16

☐ + 7 = 16

☐ + 8 = 16

☐ + 10 = 16

☐ + 13 = 16

☐ + 15 = 16

7 + ☐ = 16

☐ + 7 = 16

8 + ☐ = 16

☐ + 8 = 16

5 + ☐ = 16

☐ + 5 = 16

9 + ☐ = 16

☐ + 9 = 16

◀ Yellow Pupil Book Part 3 pages 72 and 73
▶ Copymaster Y67

Teen sums

Name ____________________

Date ____________________

Fill in the missing numbers.

9 + ☐ = 17

7 + ☐ = 17

3 + ☐ = 17

12 + ☐ = 17

4 + ☐ = 17

8 + ☐ = 17

0 + ☐ = 17

16 + ☐ = 17

5 + ☐ = 17

☐ + 2 = 17

☐ + 1 = 17

☐ + 14 = 17

☐ + 6 = 17

☐ + 17 = 17

☐ + 10 = 17

☐ + 13 = 17

☐ + 15 = 17

☐ + 11 = 17

8 + ☐ = 17

☐ + 8 = 17

6 + ☐ = 17

☐ + 6 = 17

7 + ☐ = 17

☐ + 7 = 17

9 + ☐ = 17

☐ + 9 = 17

Halves

Name ____________________

Date ____________________

Fill in the missing numbers.

Half of 26 is ________ .

Half of 56 is ________ .

Half of 90 is ________ .

Half of 106 is ________ .

Half of ________ is 19.

Half of ________ is 55.

Half of ________ is 40.

Half of ________ is 29.

How many minutes in an hour? _______ minutes.

How many minutes in an hour and a half? _______ minutes.

How many minutes in 2 hours? _______ minutes.

I'm 58.

I'm half her age.
How old am I?

____

I'm _______ .

I'm half his age.
I'm 21.

Y77

More printing

Name ____________________

Date ____________________

Fill in the missing numbers.

$9 \times 3 =$ ☐　　　　$7 \times 3 =$ ☐

$$\begin{array}{r} 15 \\ \times\ 3 \\ \hline \end{array} \quad \begin{array}{r} 25 \\ \times\ 3 \\ \hline \end{array} \quad \begin{array}{r} 35 \\ \times\ 3 \\ \hline \end{array} \quad \begin{array}{r} 30 \\ \times\ 3 \\ \hline \end{array}$$

$$\begin{array}{r} 12 \\ \times\ 3 \\ \hline \end{array} \quad \begin{array}{r} 20 \\ \times\ 3 \\ \hline \end{array} \quad \begin{array}{r} 34 \\ \times\ 3 \\ \hline \end{array} \quad \begin{array}{r} 36 \\ \times\ 3 \\ \hline \end{array}$$

$3 \overline{)66}$　　$3 \overline{)42}$　　$3 \overline{)54}$

$3 \overline{)60}$　　$3 \overline{)78}$　　$3 \overline{)93}$

$3 \overline{)84}$　　$3 \overline{)102}$　　$3 \overline{)105}$

Multiplication and division by 3 within 120 ◄ Yellow Pupil Book Part 3 pages 84 and 85 *Number Connections* © Rose Griffiths 2005

Y78

Number cards

Name ______________________

Date ______________________

Add these cards.

18 33 $\quad \begin{array}{r} 18 \\ +\ 33 \\ \hline \\ \hline \end{array}$	51 12
29 44	5 36
40 14 20 $\quad \begin{array}{r} 40 \\ 14 \\ +\ 20 \\ \hline \\ \hline \end{array}$	53 31 35
27 61 0	37 38 41
7 39 42	43 54 16

Number cards

Name ______________________

Date ______________________

Take away the smaller number from the bigger one.

23 52 $\begin{array}{r} 52 \\ -23 \\ \hline \\ \hline \end{array}$	4 28
34 22	17 26
60 48	25 57

Take away from 120.

$\begin{array}{r} 120 \\ -\ \ 18 \\ \hline \\ \hline \end{array}$ $\begin{array}{r} 120 \\ -\ \ 33 \\ \hline \\ \hline \end{array}$ $\begin{array}{r} 120 \\ -\ \ 51 \\ \hline \\ \hline \end{array}$

18 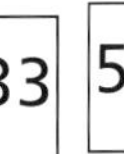
33
51 29 36 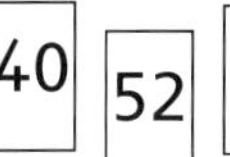
40
52
7

$\begin{array}{r} 120 \\ -\ \ 29 \\ \hline \\ \hline \end{array}$ $\begin{array}{r} 120 \\ -\ \ 36 \\ \hline \\ \hline \end{array}$ $\begin{array}{r} 120 \\ -\ \ 40 \\ \hline \\ \hline \end{array}$ $\begin{array}{r} 120 \\ -\ \ 52 \\ \hline \\ \hline \end{array}$ $\begin{array}{r} 120 \\ -\ \ 7 \\ \hline \\ \hline \end{array}$

Guinea pig sums

Name ______________________

Date ______________________

Y80

3s and 4s

Fill in the missing numbers.

3 + 3 = _____   3 + 4 = _____   4 + 4 = _____

3 + 3 + 3 + 3 + 3 + 3 = _____

4 + 4 + 4 + 4 + 4 + 4 = _____

3 + 3 + 3 + 3 + 3 + 3 + 3 + 3 + 3 = _____

4 + 4 + 4 + 4 + 4 + 4 + 4 + 4 + 4 = _____

3 + 4 + 3 + 4 + 3 + 4 = _____

3 + 3 + 4 + 3 + 3 + 4 = _____

3 + 3 + 4 + 3 + 3 + 4 + 3 + 3 + 4 = _____

Write a sum with 3s and 4s which makes 40.

30 + 30 = _____   30 + 40 = _____   40 + 40 = _____

Guinea pig sums

Name ______________________

Date ______________________

Times 3, times 4

Fill in the missing numbers.

$2 \times 3 =$ ______ $5 \times 3 =$ ______ $3 \times 8 =$ ______

$3 \times 6 =$ ______ $3 \times 0 =$ ______ $3 \times 3 =$ ______

$7 \times 3 =$ ______ $3 \times 9 =$ ______ $4 \times 3 =$ ______

$25 \times 3 =$ ______ $21 \times 3 =$ ______ $18 \times 3 =$ ______ $15 \times 3 =$ ______ $17 \times 3 =$ ______

$16 \times 3 =$ ______ $23 \times 3 =$ ______ $13 \times 3 =$ ______ $19 \times 3 =$ ______ $22 \times 3 =$ ______

$6 \times 4 =$ ______ $7 \times 4 =$ ______ $4 \times 8 =$ ______

$4 \times 4 =$ ______ $9 \times 4 =$ ______ $4 \times 10 =$ ______

$16 \times 4 =$ ______ $20 \times 4 =$ ______ $17 \times 4 =$ ______ $13 \times 4 =$ ______ $18 \times 4 =$ ______

Inches and halves

Name ____________________

Date ____________________

Y82

Make your own tape measure in inches, to help you practise using halves.

- Cut out the 3 strips and glue them together.
- Fill in the missing numbers.

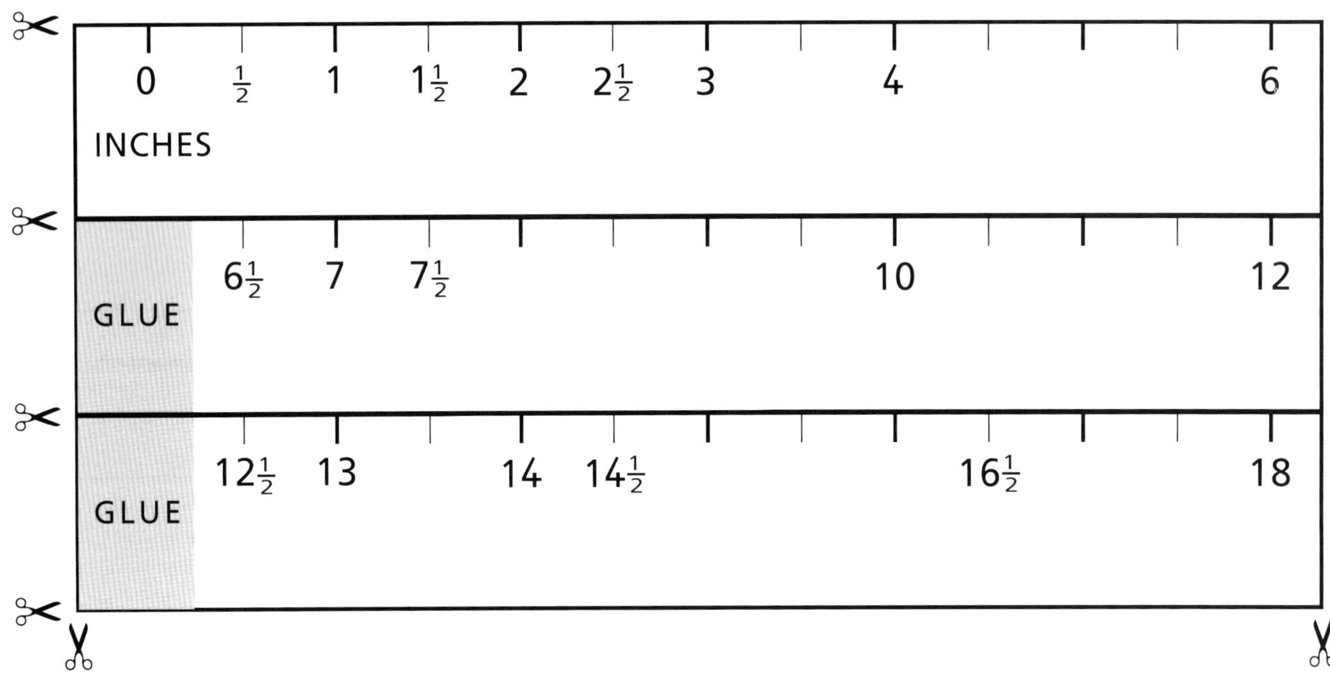

0 ½ 1 1½ 2 2½ 3 4 6
INCHES
GLUE
6½ 7 7½ 10 12
GLUE
12½ 13 14 14½ 16½ 18

Use your tape to measure things in your classroom.

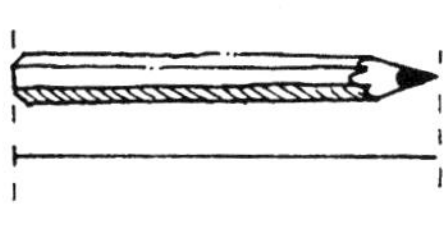

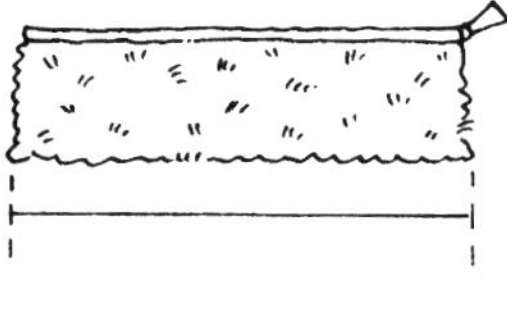

____________________ ____________________ ____________________

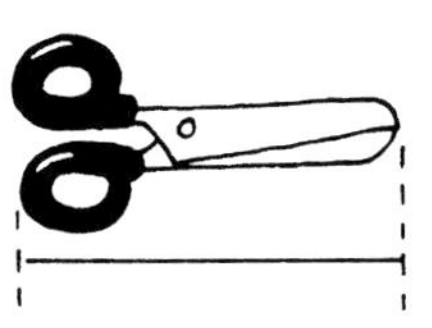

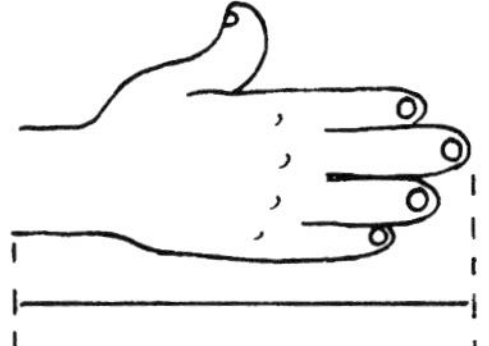

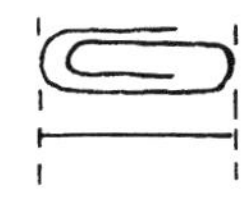

____________________ ____________________ ____________________

Y83

Inches and halves

Name ____________________

Date ____________________

How tall is each elephant? Measure to the nearest $\frac{1}{2}$".

__________ __________ __________ __________

Join the dots in order. 0, $\frac{1}{2}$, 1, $1\frac{1}{2}$, ...

Café

Name ____________________

Date ____________________

Y84
T

Make up your own price list for a café.
Choose your own prices. Choose flavours for the cans.

Your name → ______________________________'s café

Toasted sandwich	Monster muffin	Crisps
Tea	Coffee	Hot chocolate
Cans	Milkshake	

Cut out. Use with Copymaster Y85.
Ask your teacher how to play.

Halves

Name ____________________

Date ____________________

Y69

We're sharing biscuits.

If we need to,
we can break them in half.

Half of 15 is $7\frac{1}{2}$

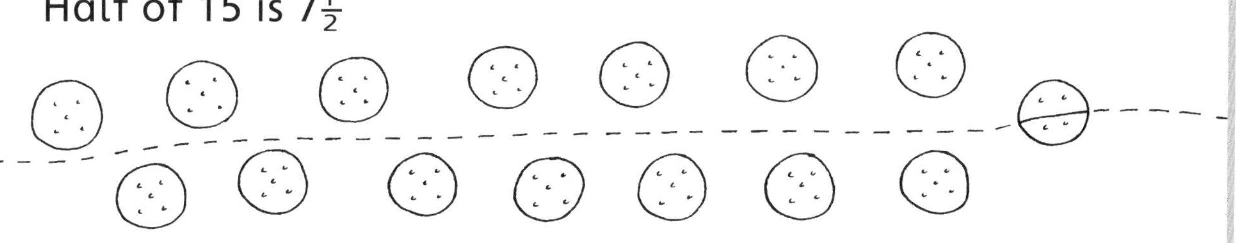

Use biscuits! (or counters) if you want to, then a calculator.

Half of 44 is ________ 4 4 ÷ 2 = ________

Half of 45 is ________ 4 5 ÷ 2 = ________

Half of 52 is ________ 5 2 ÷ 2 = ________

Half of 53 is ________ 5 3 ÷ 2 = ________

Half of 66 is ________ 6 6 ÷ 2 = ________

Half of 67 is ________ 6 7 ÷ 2 = ________

Half of 78 is ________ 7 8 ÷ 2 = ________

Half of 79 is ________ 7 9 ÷ 2 = ________

Speedy tables E

1 2 3 minute test

Name ______________________

Date ______________________

$4 \times 6 =$ ____	$14 \div 2 =$ ____	$5 \times 7 =$ ____
$8 \times 2 =$ ____	$20 \div 4 =$ ____	$18 \div 2 =$ ____
$0 \times 7 =$ ____	$24 \div 3 =$ ____	$4 \times 4 =$ ____
$3 \times 9 =$ ____	$80 \div 10 =$ ____	$45 \div 5 =$ ____
$6 \times 5 =$ ____	$15 \div 5 =$ ____	$10 \times 10 =$ ____
$7 \times 3 =$ ____	$9 \div 9 =$ ____	$18 \div 3 =$ ____
$6 \times 6 =$ ____	$27 \div 9 =$ ____	 Score: ____

Speedy tables F

1 2 3 minute test

Name ______________________

Date ______________________

$6 \times 4 =$ ____	$30 \div 5 =$ ____	$5 \times 8 =$ ____
$8 \times 3 =$ ____	$32 \div 8 =$ ____	$28 \div 4 =$ ____
$5 \times 5 =$ ____	$15 \div 3 =$ ____	$4 \times 8 =$ ____
$2 \times 0 =$ ____	$20 \div 5 =$ ____	$21 \div 3 =$ ____
$7 \times 2 =$ ____	$50 \div 10 =$ ____	$4 \times 9 =$ ____
$1 \times 9 =$ ____	$40 \div 4 =$ ____	$12 \div 2 =$ ____
$4 \times 7 =$ ____	$16 \div 2 =$ ____	 Score: ____

Printing

Name ______________________

Date ______________________

Y71

Fill in the missing numbers.

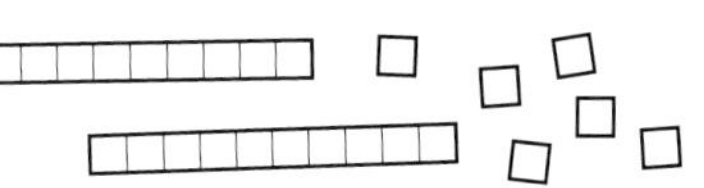

$\square \times 2 = 18$ 　　 $2 \times \square = 16$

$\begin{array}{r} 26 \\ \times\ 2 \\ \hline \end{array}$ 　 $\begin{array}{r} 31 \\ \times\ 2 \\ \hline \end{array}$ 　 $\begin{array}{r} 14 \\ \times\ 2 \\ \hline \end{array}$ 　 $\begin{array}{r} 45 \\ \times\ 2 \\ \hline \end{array}$

$\begin{array}{r} 32 \\ \times\ 2 \\ \hline \end{array}$ 　 $\begin{array}{r} 56 \\ \times\ 2 \\ \hline \end{array}$ 　 $\begin{array}{r} 27 \\ \times\ 2 \\ \hline \end{array}$ 　 $\begin{array}{r} 48 \\ \times\ 2 \\ \hline \end{array}$

$14 \div 2 = \square$ 　　 $20 \div \square = 10$

$2\overline{)26}$ 　 $2\overline{)36}$ 　 $2\overline{)46}$

$2\overline{)58}$ 　 $2\overline{)68}$ 　 $2\overline{)78}$

Y72

Printing

Name ______________________

Date ______________________

10 penguins on a rubber stamp.

How many penguins when I print ...

5 times? ________

3 times? ________

10 times? ________

8 times? ________

9 times? ________

11 times? ________

60 penguins! How many times did I print?

____

Fill in the missing numbers.

$40 \div 10 = \square$

$\square \div 10 = 2$

$70 \div 10 = \square$

$120 \div 10 = \square$

$\square \div 10 = 8$

$\square \div 10 = 11$

◀ Yellow Pupil Book Part 3 pages 78 and 79

Y73

Sums to twenty

Name ______________________

Date ______________________

Fill in the missing numbers.

5 + ☐ = 18

☐ + 7 = 17

9 + 6 = ☐

☐ + 6 = 17

7 + ☐ = 14

12 + ☐ = 19

☐ + 4 = 17

14 + 6 = ☐

15 + ☐ = 18

☐ + 10 = 20

☐ + 8 = 17

18 + ☐ = 19

Check.  ✓ or ✗

☐ + 19 = 19

☐ + 2 = 20

17 + ☐ = 20

☐ + 16 = 19

14 + 4 = ☐

☐ + 0 = 20

13 + ☐ = 19

9 + 9 = ☐

7 + ☐ = 20

☐ + 6 = 18

Cheesecakes

Name ______________________

Date ______________________

Draw lines to show how you would cut up these cheesecakes.
Use pencil not pen.

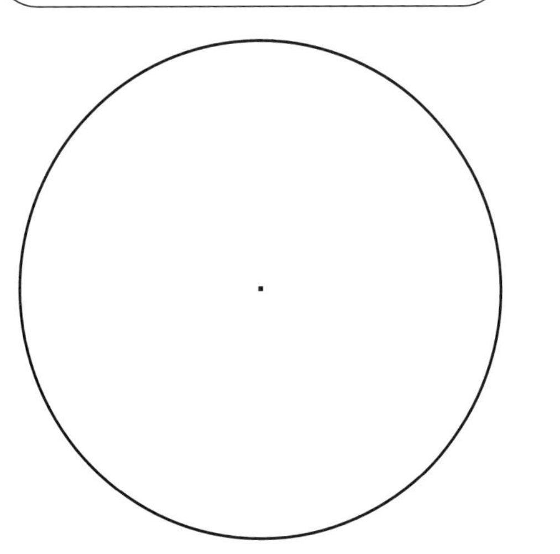
2 people ... $\frac{1}{2}$ each.

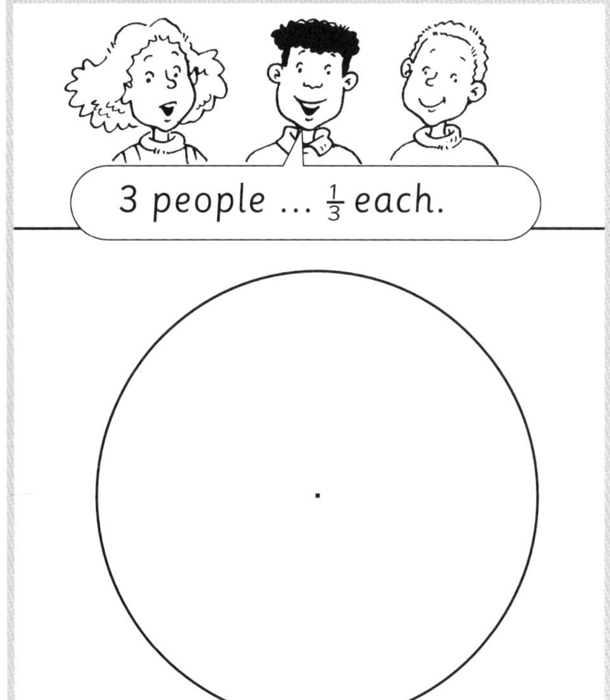
3 people ... $\frac{1}{3}$ each.

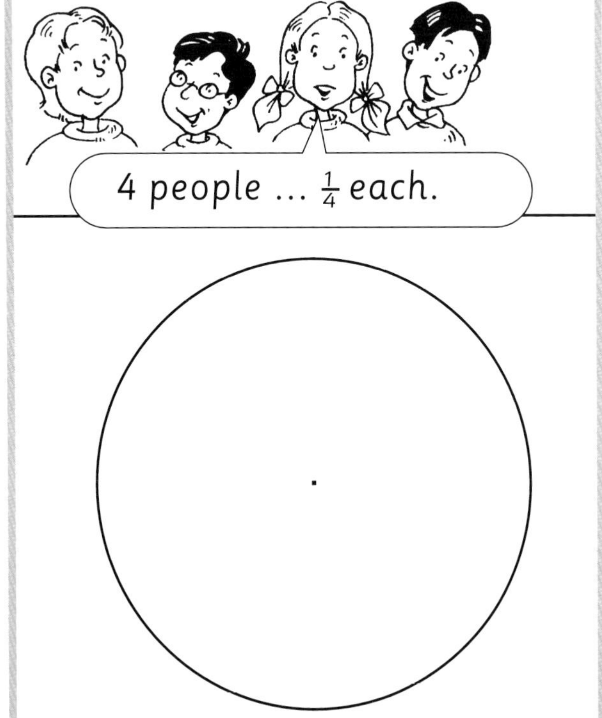
4 people ... $\frac{1}{4}$ each.

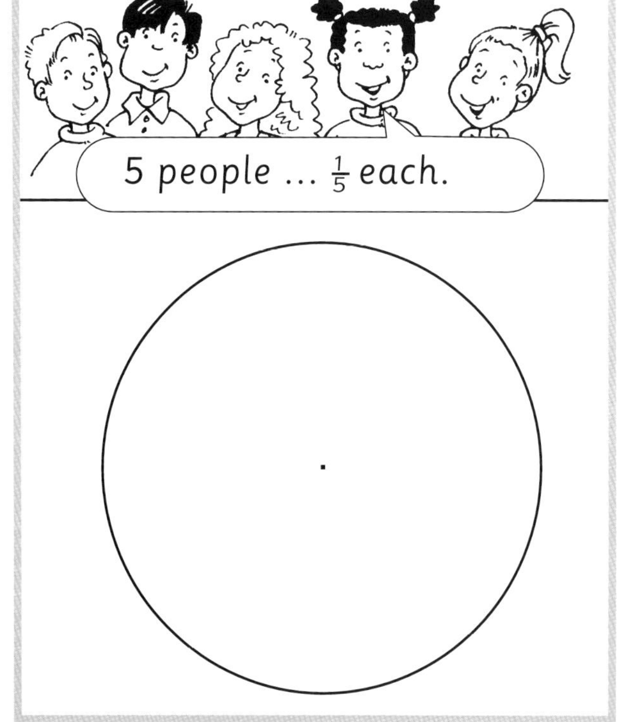
5 people ... $\frac{1}{5}$ each.

Cheesecakes

Name ____________________

Date ____________________

Draw lines to show how you would cut up these cheesecakes.

halves	thirds	quarters	fifths
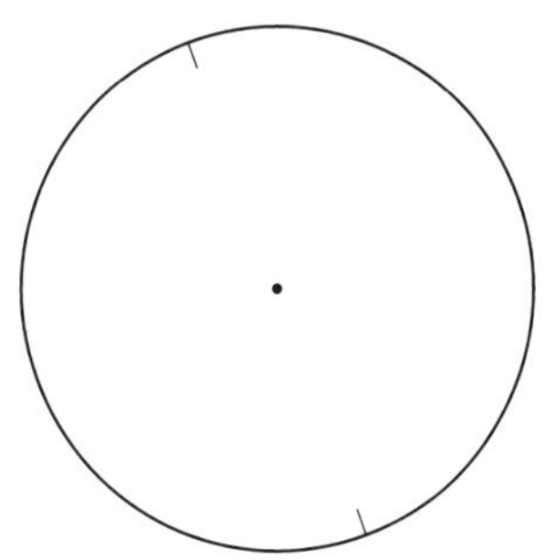	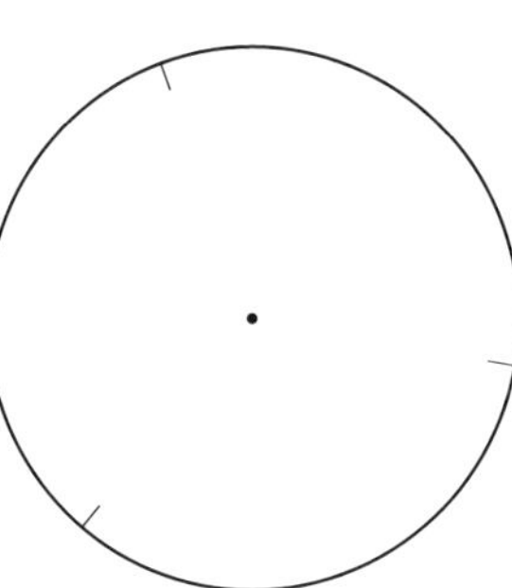	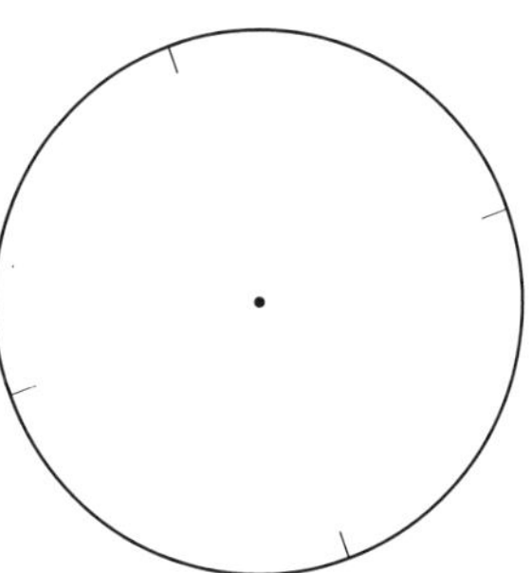	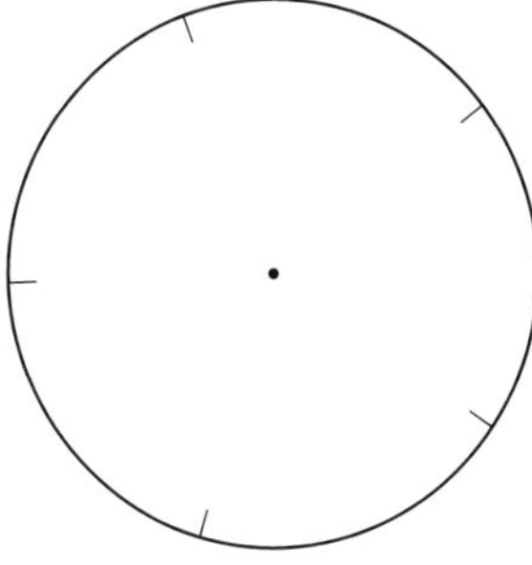

Which is bigger, $\frac{1}{2}$ or $\frac{1}{4}$? ____________

Which is bigger, $\frac{1}{2}$ or $\frac{1}{3}$? ____________

Which is bigger, $\frac{1}{4}$ or $\frac{1}{3}$? ____________

Which is bigger, $\frac{1}{3}$ or $\frac{1}{5}$? ____________

Which is bigger, $\frac{1}{5}$ or $\frac{1}{4}$? ____________

Which is bigger, $\frac{1}{2}$ or $\frac{1}{5}$?

____

Which is smaller, $\frac{1}{2}$ or $\frac{1}{3}$?

____

More printing

Name ____________________

Date ____________________

Fill in the missing numbers.

$\square \times 3 = 27$

$8 \times 3 = \square$

$$\begin{array}{r} 14 \\ \times\ 3 \\ \hline \\ \hline \end{array} \qquad \begin{array}{r} 24 \\ \times\ 3 \\ \hline \\ \hline \end{array} \qquad \begin{array}{r} 17 \\ \times\ 3 \\ \hline \\ \hline \end{array} \qquad \begin{array}{r} 27 \\ \times\ 3 \\ \hline \\ \hline \end{array}$$

$$\begin{array}{r} 18 \\ \times\ 3 \\ \hline \\ \hline \end{array} \qquad \begin{array}{r} 28 \\ \times\ 3 \\ \hline \\ \hline \end{array} \qquad \begin{array}{r} 38 \\ \times\ 3 \\ \hline \\ \hline \end{array} \qquad \begin{array}{r} 40 \\ \times\ 3 \\ \hline \\ \hline \end{array}$$

$21 \div 3 = \square$

$30 \div \square = 10$

$3\overline{)39}$ $\qquad 3\overline{)96}$ $\qquad 3\overline{)51}$

$3\overline{)45}$ $\qquad 3\overline{)81}$ $\qquad 3\overline{)57}$

◄ Yellow Pupil Book Part 3 pages 84 and 85
► Copymaster Y77

Café

Name ____________________

Date ____________________

Use with Copymaster Y84. Ask your teacher how to play.

BILL		BILL	
______	______	______	______
______	______	______	______
______	______	______	______
______	______	______	______
______	______	______	______
TOTAL:	______	TOTAL:	______
BILL		BILL	
______	______	______	______
______	______	______	______
______	______	______	______
______	______	______	______
______	______	______	______
TOTAL:	______	TOTAL:	______

Y86

Halves and quarters

Name ______________________

Date ______________________

How much cheesecake is on each tray?

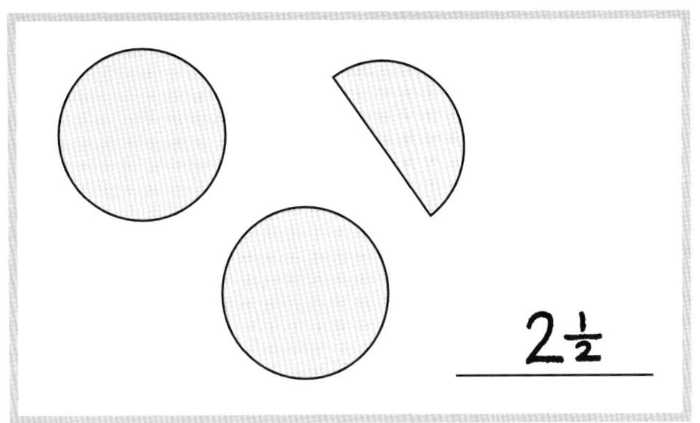
2½

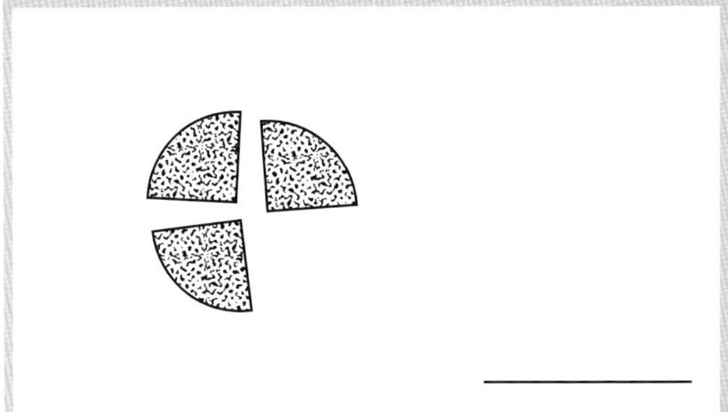

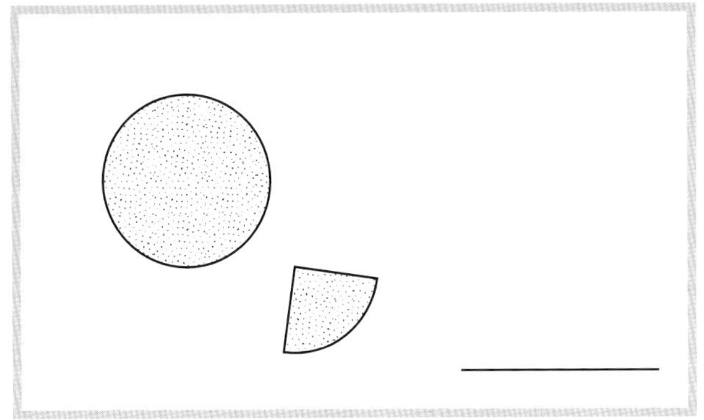

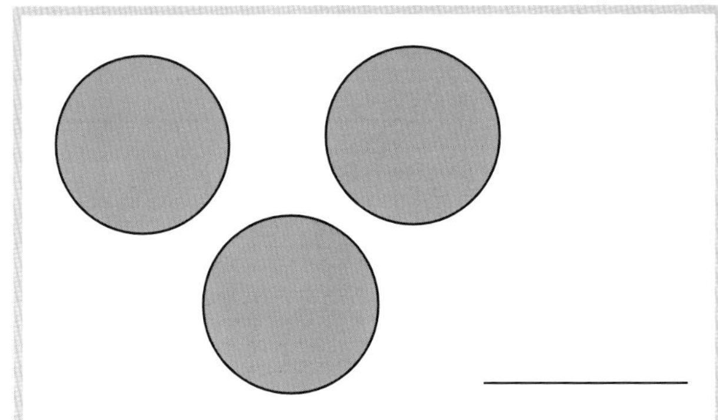

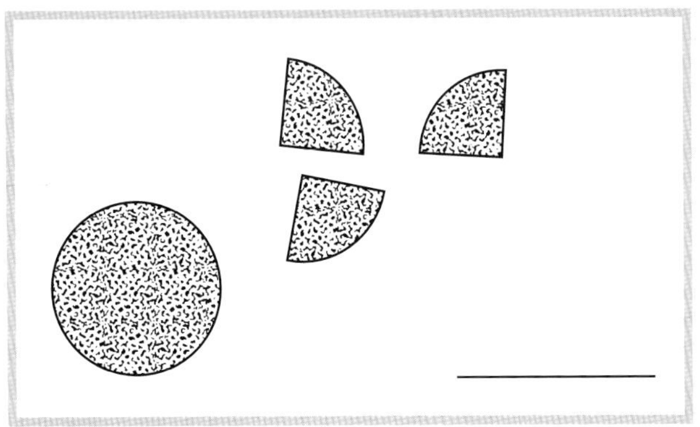

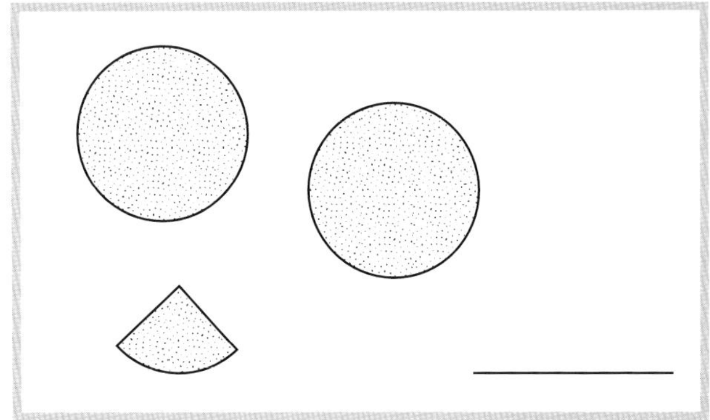

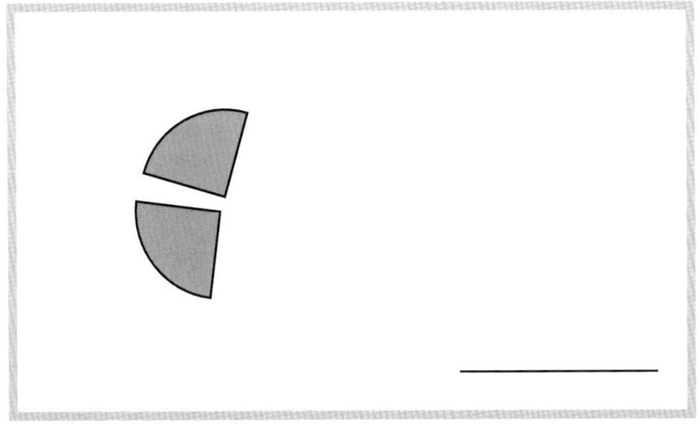

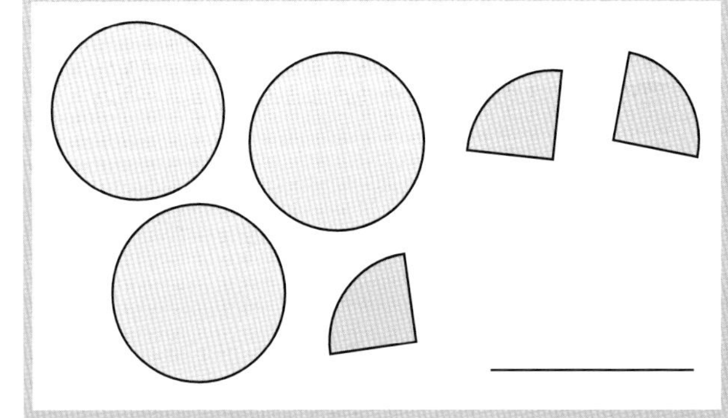

Halves and quarters

Name ______________________

Date ______________________

How long is each ribbon?
Measure to the nearest $\frac{1}{4}$".

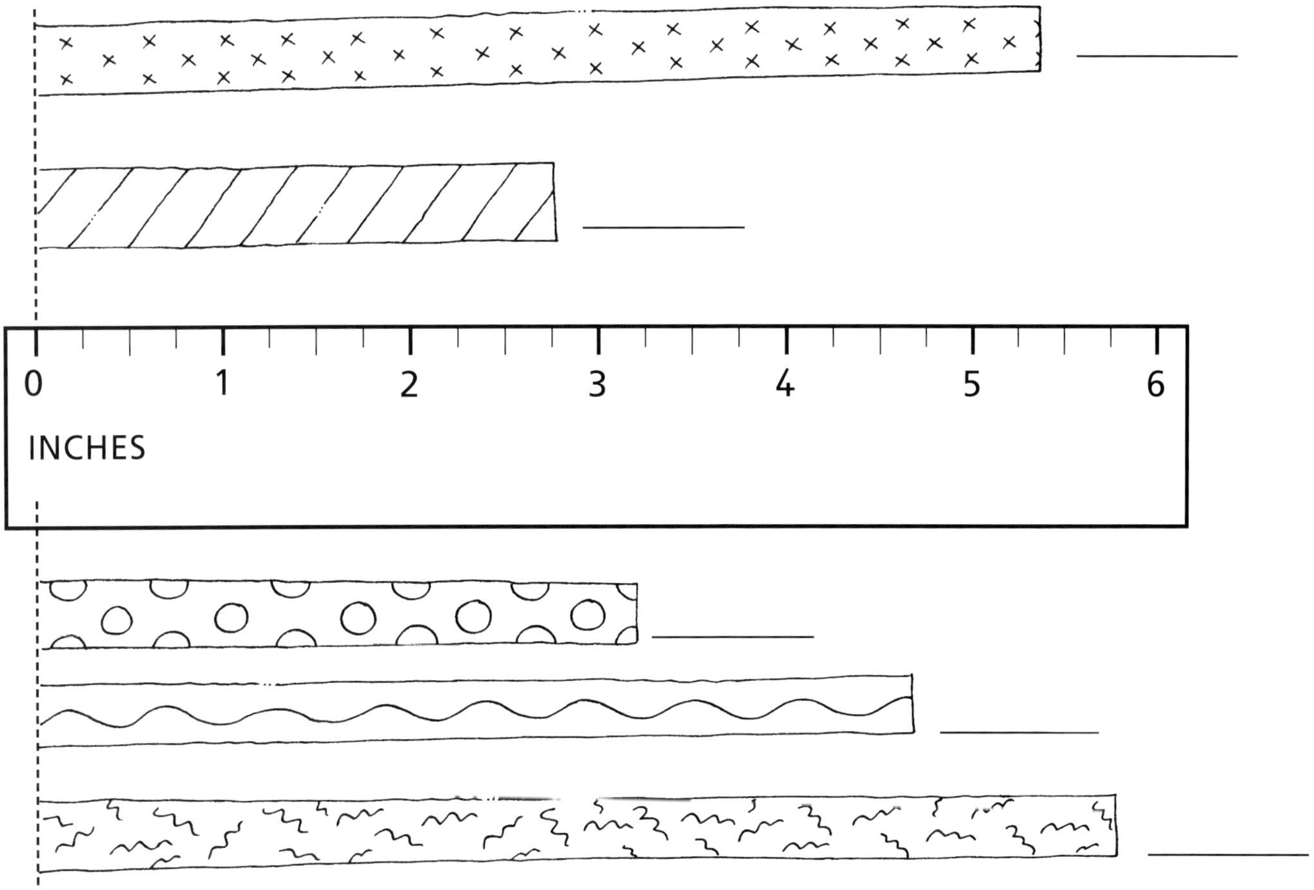

____

____

0 1 2 3 4 5 6

INCHES

____

____

____

Fill in the missing numbers.

0, $\frac{1}{2}$, 1, 1$\frac{1}{2}$, 2, 2$\frac{1}{2}$, ______, 3$\frac{1}{2}$, 4, ______, 5, ______

0, $\frac{1}{4}$, $\frac{1}{2}$, $\frac{3}{4}$, 1, 1$\frac{1}{4}$, ______, 1$\frac{3}{4}$, 2, ______, 2$\frac{1}{2}$

2, 2$\frac{1}{4}$, 2$\frac{1}{2}$, 2$\frac{3}{4}$, 3, ______, ______, ______, 4

Eighteens

sheet 1 of 2

Y88
T

GP

Print on card if possible. Reusable.
Cut out the instructions card and 20 number cards.
Store in a clear zip-top wallet or in an envelope.

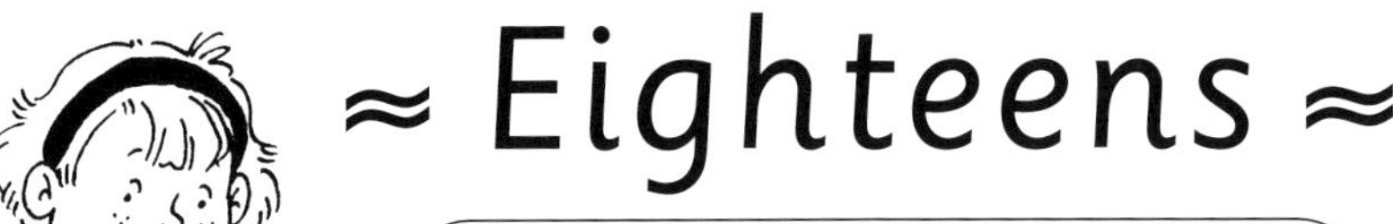
≈ Eighteens ≈
A game for 1, 2 or 3 people.

- **Before you start**
  Shuffle the number cards.
  Spread them out on the table, face down.

- **How to play**

Turn over 2 cards.
Add up the numbers.

If you get exactly 18,
keep the cards.
If not, turn the cards back over.

Now it is your friend's go.

- **Keep going until all the cards have gone.**

◀ Yellow Pupil Book Part 3; **Addition bonds to 16, 17 and 18**

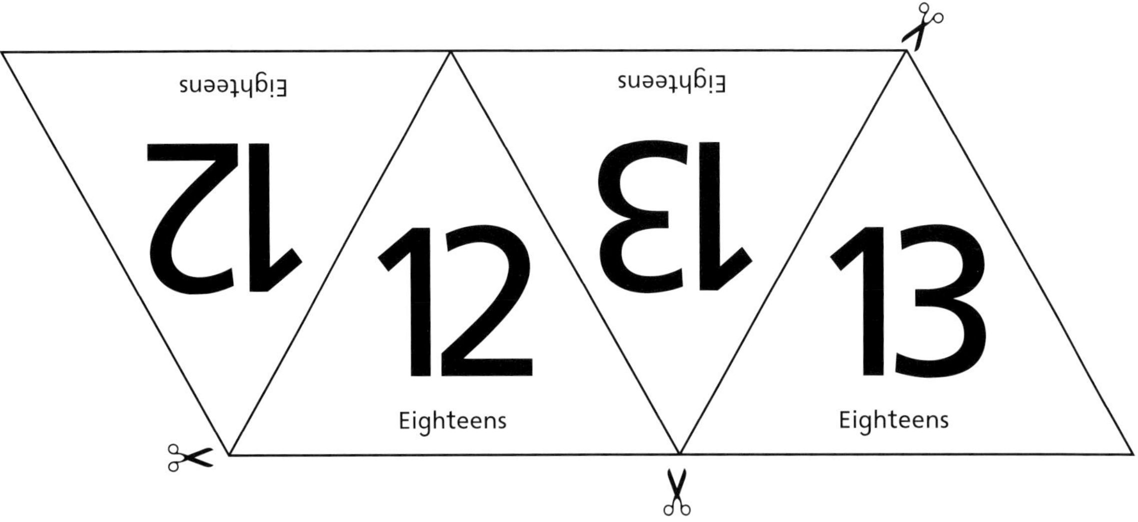
Eighteens
12
12
Eighteens
Eighteens
13
13
Eighteens

Eighteens

sheet 2 of 2

Y89
T

GP

Print on card if possible. Reusable.

Eighteens
5

5
Eighteens

Eighteens
six
9

6
six
Eighteens

8
Eighteens

Eighteens
8

7
Eighteens

Eighteens
7

Eighteens
nine
6

9
nine
Eighteens

Eighteens
nine
6

9
nine
Eighteens

11
Eighteens

Eighteens
11

10
Eighteens

Eighteens
10

Make 20

sheet 1 of 3

Print on card if possible. Reusable.
Cut out the instructions card, 3 sum cards, and 20 number cards.
Store in a clear zip-top wallet or in an envelope.

≈ Make 20 ≈
A game for 1, 2 or 3 people.

- **Before you start**
You need about 6 tens and 40 ones.
Take a sum card each.
Shuffle the number cards. Put them in a pile, face down.

- **How to play**

Take the top 2 number cards.
Put them on your sum card.
≈ Make 20 ≈
5 + 8 + =20
Then use some of the tens and ones to make 20.

≈Make 20≈
5 + 8 + =20

Now it is your friend's go.

- **Keep going until you have all filled your cards.**
Count up your tens and ones. Who collected most?

◀ Yellow Pupil Book Part 3; **Addition and subtraction bonds within 20**

Make 20

sheet 2 of 3

Y91

T

GP

Print on card if possible. Reusable.

Make 20 1	Make 20 2	Make 20 3	Make 20 4
Make 20 5	Make 20 6 six	Make 20 7	Make 20 8
Make 20 9 nine	Make 20 10	Make 20 1	Make 20 2
Make 20 3	Make 20 4	Make 20 5	Make 20 6 six
Make 20 7	Make 20 8	Make 20 9 nine	Make 20 10

Make 20

sheet 3 of 3

Print 3 copies on card if possible. Reusable.

≈ Make 20 ≈

☐	+	☐	+	= 20
☐	+	☐	+	= 20
☐	+	☐	+	= 20